A Social History of Rural Ireland in the 1950s

A Social History of Rural Ireland in the 1950s:

Remembering Crotta

By

John Galvin

Cambridge Scholars Publishing

A Social History of Rural Ireland in the 1950s: Remembering Crotta

By John Galvin

This book first published 2017

Cambridge Scholars Publishing

Lady Stephenson Library, Newcastle upon Tyne, NE6 2PA, UK

British Library Cataloguing in Publication Data
A catalogue record for this book is available from the British Library

Copyright © 2017 by John Galvin

All rights for this book reserved. No part of this book may be reproduced, stored in a retrieval system, or transmitted, in any form or by any means, electronic, mechanical, photocopying, recording or otherwise, without the prior permission of the copyright owner.

ISBN (10): 1-4438-4493-4
ISBN (13): 978-1-4438-4493-2

For my father

Table of Contents

Foreword ix
Jon Nixon

Acknowledgements xi

Prologue xiii

Chapter One 1
A Brief History of Crotta Great House

Chapter Two 13
The Past and the Present

Chapter Three 25
The Orchards

Chapter Four 31
Family History

Chapter Five 49
Our Animals

Chapter Six 69
Field Work

Chapter Seven 87
The Old and the New

Chapter Eight 95
Hurling

Chapter Nine 101
Christmas

Chapter Ten 107
The Morris Minor

Chapter Eleven 117
School

Chapter Twelve 141
Catholic Faith

Chapter Thirteen 151
The Leens

Chapter Fourteen 167
St. Brendan's

Foreword

Dr John Galvin has written a book that is both simple in its narrative unfolding and complex in its layering of remembrance and time. It appeals to the reader on many levels. Anyone interested in childhood and how childhood memories inform our adult view of the world will find this book of interest. Home – whatever and wherever that home may be – is where we start from. Our origins are grounded in a sense of home, from which we set our bearings, gain a tentative sense of direction and to which we keep returning in memory if not in lived reality. Whatever childhood means is entangled in our often complex, ambivalent and recurring feelings towards a place that we call home. His simplicity and eloquence of expression evoke not only his own remembered childhood home, but stir in the reader a recollection of childhood that resonates in the present and nudges us through into our uncertain futures

This is also a book that will be of interest to social historians, particularly those who have an interest in how the existing rural way of life adapted to the changes of modern Ireland. It is a book steeped in social history: how rurality was experienced, what community meant, how families survived, how ruined places were lost, found and lost again. It provides important clues as to how society worked, how it hung together, how it coped with its disparities and political legacy. It uses memory as a primary resource of social history and in doing so contributes to our understanding of how our social world is evolving and how the past is continually remaking the future through our understanding of it in the present.

It is a book that invites us to consider the significance of time and memory in how we understand the world and how we locate ourselves within it. The present is always at the crossroads of a remembered past and an unknowable future. From Aristotle onwards western philosophers have addressed the problem of how we live at this crossroads. Dr Galvin brings me back to where the roads meet: the present that is redolent of the past and resonates with intimations of the future. He does it not through any deep philosophising, but by drawing the reader into his story – a story which is told with irony, humour and quiet eloquence. It is a story that is personal in its distinctive voice and cadence, social in its detail and contextualisation, and strangely philosophical in its reflections on how the past resonates in the present.

No one could fail to be moved by this evocative account of the ordinariness of a childhood spent in such extraordinary circumstances.

Jon Nixon
Honorary Professor, Centre for Lifelong Learning,
Research and Development, Education University of Hong Kong

ACKNOWLEDGEMENTS

I am extremely grateful to those who helped me bring this book to fruition:

To my wife and family for their patience and encouragement;

To my son John who sensitively reviewed the first draft, scanned and collated the centrefold, and designed the front cover;

To my son Michael who proofread the final draft;

To Professor Jon Nixon, my former supervisor at Sheffield University, whose support, guidance and encouragement helped me over the finish line.

To the prepress team at Cambridge Scholars Publishing who assisted with the final details of the text.

To NLI and RIC for the use of their photographs.

Prologue

For years I had toyed with the idea of writing a book about Crotta and the old Great House. I wanted to record something of its history and also its legacy to us, the Galvin family, who grew up amidst its ruins and its folklore. Like the Great House, our way of life in rural Ireland of the mid nineteen hundreds has largely disappeared but the memories live on and none remain so vivid as those of early childhood - those carefree times when wonder and awe were ever present and everyday was a surprise package to be opened and enjoyed, or sometimes just endured.

But my life was far too busy to ever get around to writing about such things until one day Fate played a hand and gave me all the time in the world. It was a lovely June afternoon when Death came a-calling but certain fortuitous circumstances prevented Death from taking me. Michael had come to help out 'his old man' on a job I was doing and afterwards enjoy a barbeque in the back garden. I was laying paving slabs underneath the clothes line, cutting them to shape with an angle-grinder. "Give me ten minutes" I shouted, "I'll finish this little bit and we'll have a beer". That 'little bit' was to have dramatic consequences and send my life, as I knew it, into a tail-spin.

As I was cutting the last slab, the disk shattered sending a piece of rotating shrapnel up along my arm, butchering it from wrist to shoulder. The rotating piece of metal waltzed its way through the flesh of my left arm severing nerves, tendons, muscle and arteries. I staggered towards the backdoor in shock and horror, blood spilling on the ground. Angela reacted quickly. Pale-faced and stunned she tightly wrapped a towel around the bloodied shirtsleeve and screamed at Michael to start the car. I knew it was bad. I had glanced in horror at the cut and the amount of blood on the ground told a frightening story.

Michael speeds off towards St. John's hospital like a bat out of hell. Somewhere along the way my vision begins to dim and then blanks out. I'm aware of the sound of the horn and Michael cursing profusely as he tries to make time through traffic lights and congested streets. We make it to St John's A & E in ten minutes and because I have fallen sideways on the back seat the blood loss is impeded and some vision has returned. The sight of my blood-drenched clothes and bloodied car seat is more than I can handle. I feel sick. Michael helps me out of the car and in the midst of my horror and panic I hear myself saying "You're parked on double yellow lines", as he rushes me through the emergency door.

Luckily St. John's A & E was still open; it was now about 5.30 p.m. and the A & E service closes at 6.00. The only other hospital is situated at the other side of the city. I would not have made it across town, certainly not at rush hour. My candle would have gone out somewhere along O'Connell Avenue as the last few pints of blood drained out through my ruptured arm. So 'thank you God' and the forces that combined to have me on the emergency table at St. John's in the nick of time. An excellent emergency team kicked into action immediately.

As I lie there shuddering and groaning, I can hear the scissors cutting through my clothes and I'm acutely aware of the urgency in the voice of a male doctor who is barking out commands with a foreign accent. His instructions are clipped and precise, so many mils of this and that and he makes constant enquiry with regard to blood pressure. While some kind female nurse or doctor keeps saying to me, "You're okay pet, you're going to be alright", I can feel the strange discomfort of a catheter being inserted into my penis and like a small boy needing to be comforted I continue moaning and groaning long after they have stabilised my blood pressure and ended my pain.

St. John's emergency team had done their work well. My life was saved but my left arm was now without a blood supply so I was taken immediately by ambulance to the Mid-West Regional Hospital, as it was known then. Michael travelled with me in the ambulance, leaving his blood drenched car abandoned on the double yellow lines. I remember making vague bits of conversation with the accompanying nurse as frightening thoughts race through my mind. The world I loved so much was now falling apart. I had always been a DIY man – the garden was my playground, designed and re-designed to suit my fancy. What did the future hold for a one-armed DIY man? These were my worries. After what seemed like an endless wait in the Mid-West Regional Hospital, I was wheeled to the operating theatre. As I was being rolled in the anaesthetist whispered in my ear that I should be prepared for the worst; that the arm might be beyond saving. I quickly interjected saying, "Don't let my wife hear that," not knowing that Angela was right behind the trolley. In a little while a mask was put over my mouth and nose and 'the lights went out'.

Professor Grace saved my arm. During three hours of micro-surgery he removed three veins from my left thigh and patched up the lacerated arteries in my arm. The plumbing was then complete and the blood flow could circulate once more due to the wonders of medical science and the gifted hands of the surgeon and his team. But the wiring, the plastics, could not be done in Limerick so the following day I was taken to Cork University Hospital. It was six o'clock when we got through the entry procedure at CUH. They parked my trolley along the main corridor where I would remain for the night with my left arm suspended on a steel pole. My family gathered round me and we talked for a long time.

I'm trying to be upbeat and blasé, making jokes and wisecracks. I'm putting on some kind of show in response to Angela's drawn expression as she looks at me with frightened eyes. Eventually I insist they go home and I try to make the best of my situation. I have been told to remain fasting from midnight for tomorrow's operation but I haven't eaten in twenty four hours so I ask if I can have a cup of tea and two slices of bread. I wait patiently. I think they have forgotten. I can see that the nurses are run off their feet. Eventually a staff member places a tray on my lap, but is gone before I can ask her to butter the bread. I swallow the unbuttered bread along with my pride and frustration as I succumb to a growing sense of depression and despair. But sleep comes to the rescue. In spite of my arm being slung up on the steel pole and a drainage bag attached and the catheter causing great discomfort every time I move, I drift off to sleep.

Early the following morning I was wheeled into a vacant cubicle in Ward 3B. Mr. O Shaughnessy, the plastics consultant, and his team came to visit me and informed me that I would be going to surgery at two o'clock. I waited anxiously, my mind in a whirl. The time went by slowly and then I was wheeled along winding corridors to the operating theatre. The theatre staff were gentle and re-assuring and trying hard to be cheerful. The anaesthetist put a mask over my face. I smiled a goodbye and the lights went out again.

Three hours later they are trying to bring me around. I don't want to wake up. I feel so cosy and warm in my morphine haze. In my groggy state I'm taken to Ward 3B and later that evening Mr. O'Shaughnessy and his team come to see me again. They tell me that this part of the surgery has been successful. Tendons and muscles have been stitched back together but my median nerve is in a bad state and that it was difficult to say if anything worthwhile could be done. They explain that a nerve graft (taking nerve strands from my leg and grafting them into my arm) might be possible, but don't hold out much hope. My heart sinks and I mutter something about being a good fighter and ready to do whatever it takes. Mr O'Shaughnessy smiles benignly and points out that nerve growth, for older people, is extremely slow. The team moves on to another patient. I lie there in a dazed state until the ringing of my mobile brings me back. It's Diarmuid, my eldest son, enquiring how I am. In a despairing tone I relay the doctor's report to him. His advice is instant and decisive; that having the nerve graft done would at least provide my arm with the 'hardware' for the body to work with over the coming years. He tells me I should demand that the third operation go ahead. This helps but I'm in such a depressed state that I request a sleeping tablet from the night nurse. The following day, Mr. Kelly, another hand specialist, comes to visit me. He's gentle and patient and a good listener. He offers further hope, suggesting that a nerve transfer onto the ulnar nerve line, which carries the nerve signal to the 'ring' and 'little' fingers, might be possible. He would have to confer with another plastics specialist in Dublin and would let me know on Monday. I spend the weekend in an agitated state, moving from prayer and hope to self-recrimination with a continuous litany of regretful 'if only'.

CUH is a training hospital so being a guinea pig is par for the course. Trainee nurses kept coming at frequent intervals to check my blood pressure, my temperature and the drainage bag on my arm. My arm was fixed in a 90 degree brace which had to remain upright at all times so the oozing could drip into the bag. The night nurse, Sara, was very gentle when administering the intra-venous antibiotic injections. Foregoing my embarrassment I asked if she could remove the catheter. Removing it was a bit painful but I was delighted with my new sense of freedom. I could now walk around more comfortably and explore my surroundings. I discovered a very pleasant coffee dock area - a high-roofed glass enclosure with potted trees and plants giving the impression of being outdoors. It didn't smell of hospital. This was where I would spend much of my time.

I had just got back to my bed on that Saturday evening when my mobile rang. It was my brother Diarmuid with the devastating news that my mother had died. She was ninety-six and had been growing steadily weaker in recent months. I had meant to get down to see her but now I would not even be at her funeral. She had slipped away quietly in her sleep; she was never one to make a fuss. The wake would be on Monday and the burial on Tuesday and I would not be there. The nurses and other patients were very sympathetic but my heart felt like a lump of lead in my chest. My mind kept wandering back to sunny afternoons as she and I strolled along Banna strand or Ballyheigue beach, her bony fingers gripping my elbow like a vice-grip as she expressed her joy at the wonders of nature and the beauty of the world.

"Goodbye Mammy! May perpetual light shine upon you! You will live on in my memory. I will look for you around the walls and fields of Crotta. I will see you in the blossoming flowers of spring and in the ripening fruits of autumn. I will feel your presence in the warm glow of winter fires. I will hear your voice in the whispering trees, in the lowing of cows on summer evenings, in the chorus of birdsong at early dawn and in the rolling waves of Banna beach. Rest in peace! You were a good mother."

The third operation took place on the day of the funeral. The surgery team removed three strands of sensation nerves from the calf of my right leg and grafted them on to the mangled median nerve in my arm. It was late in the evening when Angela, Michael and John visited me and even though I was still a bit groggy from the anaesthetic I was delighted to get a detailed account of the funeral events. Mammy had got 'a great send-off'. I felt sure she enjoyed it from up above. She had always been a very sociable person and nothing pleased her so much as meeting friends and relations and updating her data base with all the events in their lives.

After they've gone I lapse into a dreamy nostalgia. Memories from the distant past come flooding in, images of childhood events when Mammy, as a vibrant young woman is going about her work in the house and in the farmyard. The memory of one particular day takes over, a summer's day when I'm hanging around her, following her here and there like a pet lamb. She asks if I have got anything better to do as my attempts at helping are slowing her down. She is singing softly to herself a verse from 'Carrig Donn':

"*Soft April showers and bright May flowers will bring the summer back again, but will they bring me back the hours I spent with my brave Donal then.*"

I ask her who 'Donal' was and she laughs and says just somebody in the song, but she likes that verse very much and keeps singing it over and over again.

My reverie was broken when the nurse on duty for the night came to check my temperature and blood pressure for the umpteenth time. My suspended left hand was filling up the fluid bag at an alarming rate and now, to add to my anxieties, my right leg was numb from the knee down since the last operation. But I was on a morphine drip which I could self-administer and that took the edge off my worries. The nurse brought me two 'pigeons' for the night; that's what she called the disposable urine containers. Peeing from a horizontal position was new to me and my dead right leg was uncooperative, which made it more difficult.

When I awake the following morning my right leg is still totally numb. This worries me a little because it should have come right by now. It feels so strange. I rub it against my left foot and it feels heavy and awkward and foreign. It feels like my foot is wearing a cold marble cast. When I click both insteps together it feels like tapping on a heavy metal pipe. This was not in the plan. I was told to expect a lack of sensation in the calf of my leg where the nerve strands were removed but not total numbness from the knee down. A frightening thought runs through my mind; had they accidently severed some vital connections in my leg? I banish that thought and console myself with the fact that all the operations are now over and that this is Day One of my recovery. Later that morning the surgery team, Drs. Shirley and Jeffrey, come to see how I am and inform me that the operation has been a success. I tell them about my dead leg and, quite casually, they inform me that there had been an 'incident' during the operation, which had to be reported to head office. The word 'incident' was a euphemism for 'cock-up'. In order to extract the nerve strands from the calf of my right leg, it was necessary to put on a tourniquet above the knee. The timer on the tourniquet was set to release pressure after two hours (maximum) and now they're informing me that the timer must have been faulty because it had stayed on, unchecked, for the full duration of my operation - well over four hours. No sympathy or word of sorrow is expressed - it being nobody's fault, the machine being faulty. I can feel my anger rising. I ask how long it will take for my leg to come right but they can't say as this has never happened before. When pushed for an answer Dr. Shirley says it could take one week or maybe ten weeks or maybe longer. They couldn't predict the outcome. And on that depressing note, they leave.

I lie there dumbfounded for a long time. In desperation I ring my son Michael who has been practising natural healing therapies for more than a decade. His speciality is cranio-sacral therapy, which assists and accelerates the body's healing processes. He is very confident that he can bring my leg back and his re-assurance calms me. He will come the following day and do a cranio-sacral session with me. As I lie there in my muddled state of mind, Angela arrives. She had taken the bus down and got a taxi to the hospital. I try to tell her to space out her visits but she makes little of the journey and the upset and disruption of the past week.

She reads the depression and anger on my face and offers whatever consolation she can muster. I'm now reduced to one arm and one leg. Bathroom functions have become a nightmare with serious loss of independence and dignity. On the bright side, I have a window cubicle with a pleasant view of green fields and a wooded hillside that reminds me of childhood days in Crotta. Scenes from boyhood years are filling my mind. I slip into reverie and it helps pass the time.

Mentally, I was surviving on a diet of DVDs. My little DVD player had become my cherished companion. My fellow patients must have thought me very unsociable and in truth I was. Much of my time was spent watching films within the confines of my curtained-off cubicle. In the adjacent cubicle, the tortured breathing of a very ill man kept me awake through nights of endless torture. My only refuge and comfort was to lose myself in the world of Hollywood make-believe. I watched the *Shawshank Redemption* two nights running and drew some solace from Andy Dufresne's ability to triumph over adversity. "Sweet are the uses of adversity" states the deposed duke in *As You Like It*, but I was at a loss to see any sweetness in my adversity. Was the life I loved so much now gone? The future looked bleak. I didn't want to think about the long journey back and whether or not I would get there.

The night nurse going off duty comes to administer another antibiotic injection. The access vein in my forearm collapsed yesterday and they had to put a new access aperture on the back of my hand. This hurts when they stick the syringe into it, except when Anne does it. Anne is very pretty and it's worth the pain just to have her hold my hand. I keep hoping she will remain on duty. Her positive attitude is giving me hope. She encourages me to stand on my dead leg and to keep moving it, while I hold on to the steel pole on which I hang my left arm. The hospital physiotherapist comes to have a look at me and suggests getting used to using the single crutch, which she places by my bedside. I ask her the same question I had asked the doctors. She gives the same answer; she couldn't be sure but felt it could take quite a long time. By now my entire focus is switched to my banjaxed leg. My poor traumatised arm now seems less urgent. It hangs in front of me, heavily bandaged and secured in a right-angle splint which seems to weigh a ton. My swollen fingers look like jumbo sausages and are sticking out but unable to move. Luckily, I can hobble to the bathroom because my cubicle is near the bathroom door but using the facilities is a nightmare.

So, encouraged by Anne, I stood by the steel pole next morning and counted out one hundred steps, moving my dead foot up and down. I found this exercise very therapeutic. The monotony brought a certain peace of mind. I was just back in bed when Michael and Colm arrived. They drew the curtains around my bed and after a brief conversation Michael began his magic. So I closed my eyes and gave myself up to cranio-sacral healing. He was very positive and re-assuring and explained that the healing processes would continue to integrate all through the night. I lay awake waiting for the miracle to happen, hoping for some small hint of feeling below the knee, some sensation of pain would be so welcome, but nothing happened. The consultant called in the following morning doing his rounds, accompanied by five interns. One of them changed the dressing on my leg where the nerve strands had been removed and said it was healing nicely, but I wasn't interested in the scar, I wanted to know when my leg was coming back to life. It was four days since the operation in which the malfunctioning tourniquet had killed my leg. I asked again how long it was going to take. I got an unsympathetic "We'll have to wait and see", from the surgeon.

Colm has given me a thick copybook and a good pen and because I have all the time in the world, I begin to write. I get hooked almost immediately. I can't seem to stop. My scribble is flowing freely across the pages. My brain is spilling out every event of the past week in minute detail. Not only is it a great pastime, it is also strangely energising and liberating. Eventually, I decide to take a break and do my hundred steps. I grip the steel pole with my good hand and start counting, right leg rising and falling with neither sensation nor control. I'm about halfway through when it happens. My right foot experiences an internal explosion of searing pain. I collapse in agony on the bed. The pain is excruciating as myriads of nerve ends re-awaken simultaneously but I know this is good pain. I can feel 'the red-hot pins and needles' throbbing and pulsating right down to my toes. Tears roll down my face, tears of joy coupled with tears of pain. Pain! Pain! Beautiful Pain! Thank you Pain! Thank you God! Thank you Michael! The miracle has happened. I lie on the bed for a long time afraid to put my leg to the test. Eventually I pluck up the courage to do another hundred steps. Luckily I'm facing the wall and nobody can see my face as I laugh and cry and bite on my lip to contain myself. My leg is at least 50% back, if not more. Now there's hope. I lose count of my stepping but I must have done at least a 1000 before sitting down. I draw the curtains around my bed as gratitude overwhelms me.

In the afternoon, Dr. Raz, a female Indian doctor, came to change the dressing on my arm. She opened up the layers of bandaging and there it was, my poor butchered arm, looking like a large stuffed pork steak that had been badly sown together. It was very swollen and the long snake-like cut looked red and raw. I turned my head away but she assured me it was healing nicely. She was very friendly and tried to comfort me, telling me that everything that happens to us has a purpose and that I would be better for it and that I must learn to be patient. I tried to take this on board and responded with the usual platitudes of 'that's life' and 'life has its ups and downs' but somehow, what she said registered with me. I began to see my pain as a turning point and, without realising it, the process of coming to terms with my situation had begun.

"Muileann muilte Dé go mall ach muileann siad go mion" (*God's mills grind slowly but grind exceedingly fine*).

On day fifteen, without a word of warning, the head nurse said I could be discharged just as soon as the occupational therapist fitted a new lightweight splint that would keep my elbow at a 90 degree angle to prevent rupturing the repair work. My brothers Tom, Diarmuid and Pat rang to say they were coming to visit me. I told them I was being discharged but they said they had everything planned and were looking forward to a 'day out', so I said we'd meet in the Wilton bar across from the hospital as soon as I was free. I rang home and Angela and Colm came to collect me, bringing a big loose-fitting jacket. This was my first attempt at getting into everyday clothes and it made me aware of how incapacitated I was. A whole range of little things were now beyond me, one being the tying of shoe laces. Colm went on ahead to bring the car to the hospital entrance. Angela packed my few belongings into a bag while I said goodbye to some nurses and patients; I had become more sociable towards the end. My walk was a bit unsteady so I proceeded slowly down the long corridor. My left arm was protruding out in front of me as if I were carrying a shotgun and my right leg was doing a drunken walk but in my head I was leaping and jumping for joy. I was on the mend and I was going home. We had a good laugh in the Wilton bar. It was so great to be there and hear their funny stories; stories about things that happened at the funeral and people who had asked about me. I would have loved to try a pint but I was still on an anti-biotic. As they left they told me we would be meeting again for our mother's 'month's mind' mass and we would have more time and I'd be in better shape. I finished my glass of orange and we headed for home.

A special homecoming dinner is prepared and Angela cuts the meat into bite-size portions. With a fork in my right hand, my left elbow resting on the table and my arm sticking up in the air, I wolf down the delicious roast beef and roast potatoes. I go to bed early; my energy and stamina levels are very low. I awake early next morning with the hospital routine still programmed in. In the bathroom, I set about re-learning how to wash and shave. It takes so long; everything now moves at a slower pace. I sit in the kitchen and sip a mug of tea, gazing out the window. The morning is bright and sunny so I decide to take a stroll around the back garden to see how the flowers and vegetables are doing and also to visit the scene of the accident. A shudder runs down my spine; there are still some dark patches on the pavement. I hurry back indoors and begin to set up a writing desk in a small spare room that catches the morning sun. I sit and gaze out the window as if seeing it for the first time. The array of flowers is a sight to behold; arranged by colour and height by my flower-obsessed wife. Angela spends her free time imposing her will on Nature and what a wonderful display she has produced. Beyond the flowers I can see the apple tree and the plum tree laden with new fruit and I say to myself: "This isn't too bad; things could be worse, things could easily be far worse". I know that now I have to play a waiting game. Nerve regeneration is extremely slow; the consultant said a millimetre a day. My median nerve has to regenerate down the full length of my arm so I calculate, optimistically, three years. In the meantime I'm conscious of having many months of painful rehabilitation and physiotherapy ahead but no matter, I'm on the way back. I open my copybook and start to write.

Chapter One

A Brief History of Crotta Great House

I grew up in a three-hundred year old building that was once attached to the Great House at Crotta. The house was completed in 1669 and was the seat of the Ponsonby family for almost two hundred years. Later occupants included the Kitchener family from 1852 to 1863. Lord Kitchener, Britain's highly decorated military genius, spent his boyhood years in Crotta and in 1910 came back to see the place of his youth. Six years later, he would enter a watery grave in the North Sea when the HMS Hampshire struck a mine on its way to Russia. The story goes that an officer on board attempted to save him shouting: *"Make way for Lord Kitchener"*. This story was relayed to us as children by Sara, a much loved widowed aunt who lived in the house with us. Sara used the phrase *"Make Way for Lord Kitchener"* whenever she was passing by and it became our catch-phrase when someone was blocking our way. As children, echoes of the past pervaded our everyday lives. We played hide-and-seek amidst the old ruined walls where seven generations of the Ponsonby family had enjoyed their privileged existence. While no sightings of any ghosts were ever recorded, the towering walls had a ghostly feeling about them and on dark winter nights were best avoided.

Our family home was that part of the old Great House that had survived after the rest of the building had been damaged by fire in 1920 and then slowly demolished over the coming decades. A story relates that the Black and Tans set fire to the house but on being told that they were burning Lord Kitchener's early home they immediately tried to put it out. But the roof had been seriously damaged and so bit by bit it was reduced down to a gaunt shell. Such was the ignominious end to this landmark building that had cast a giant shadow on the surrounding countryside for almost three hundred years. In my childhood day-dreams I would try to visualise this decaying ruin in its former glory and hoped that someday the remaining part might be restored, but the term 'preservation order' had not yet been coined and the last remnant of Crotta Great House slowly disappeared from the landscape.

Chapter One

Crotta is a small town-land in the parish of Kilflynn, about ten miles north of Tralee in county Kerry. The name Crotta comes from the Gaelic word *cruit* (the genitive case being *cruite*) meaning 'hump'. The 'hump' or hill in question is Neenan's Hill that lies at the southern side of what was once Crotta estate. Three Celtic ring forts in close proximity hold the secrets of its prehistoric past and its early Celtic inhabitants and tell us that this was a favoured dwelling place through the centuries. The estate at Crotta, in the barony of Clanmaurice, was granted to Captain Henry Ponsonby in 1666 under the Acts of Settlement as his reward for faithful service to Cromwell in the conquest of Ireland. It remained in Ponsonby hands until after the Great Famine when it was put up for sale by Thomas Carrique Ponsonby through the Encumbered Estates Act. The bankrupt estate was then held in lease from the Court of Chancery by Samuel Julian Esq. and occupied by his steward who leased it to Colonel H. Kitchener, a retired army officer of an Indian regiment. The Kitchener family lived there for twelve years or so before moving to Switzerland in 1863. The estate then came into the possession of the Browne family, who were also descendants of Cromwellian officers that had acquired large estates in Ventry, Kenmare and Tarbert. Some fourteen years later, Crotta was once again put up for sale by Thomas Beale Browne in 1877. At the end of the nineteenth century the estate was owned by Richard Savage and in 1925 it was bought by two Carmody brothers who divided the remaining part of the estate and the Great House into two family farms and dwellings. One of these farms and the remaining part of the 'old house' became the Galvin family home in 1940. This was where I grew up. This was 'our home' and remained so until our mother, its last occupant, moved into a 'granny flat' in 1991.

From an early age, I was fascinated by the history and stories of the old Great House. As children we played amidst its dilapidated walls and its ruined grandeur filled us with wonder and awe. It influenced our thinking and impacted on our lives down through the years as we tried to unlock its secrets and piece together its buried history. The house had been a fine Elizabethan mansion of two storeys with five tall chimneys. The front of the house had two projecting wings with high gables on which could be seen the Ponsonby coat of arms. Thirteen large windows and one lunette window over the entrance porch looked out onto green pasture land, which still carries the name 'the front lawn'. The rear view of the house was equally impressive with a magnificent arched entrance within which the staircase was housed. The big green field at the back of the house is still referred to as 'the back lawn'. Attached to the northern side of the house was a slightly lower building that housed the household staff and attached onto this was a coach house and stables. At the back of the stables was the Pound in which the carriage horses were confined when transport was needed. Much of the estate was covered with great oak woods, the remnants of which remain on the hillside that gave Crotta its name.

Because of its direct link to Crotta, the Cromwellian conquest of Ireland was of special interest to me in history lessons at school. These lessons were implanted in our young minds with stirring accounts of the injustice done. Cromwell, the arch villain of the story, arrived in Ireland in 1649 with an army of 12,000 battle-hardened soldiers. They were fully equipped with the most advanced weaponry and well provided for with supplies from England. Cromwell, "Like a lightning passed through the land". His campaign, which was accompanied by blood, carnage and destruction, had a triple purpose: Firstly, to avenge the reported massacre of the Ulster Protestant planters, secondly, to destroy Catholicism in Ireland and thirdly, to take over as much land as possible, in order to pay those who had advanced money to the English parliament in exchange for promises of land in Ireland and those who had volunteered to fight. Captain Henry Ponsonby, the future owner of Crotta estate, was one such adventurer. The previous owners, the deStack family of Norman descent, were forced out. During the conquest of north Kerry their residence at Crotta was used as a barracks by the Cromwellian army. After a campaign of barbaric cruelty, the defeated Irish and Anglo-Irish were forced to leave their lands and move west of the Shannon. This mass exodus is summed up in a saying that has lived on in folk memory: *'To Hell or to Connaught'*. While tens of thousands obeyed the order many more remained behind as Tories (outlaws), hiding out in remote regions, waiting for the opportunity to strike back.

By 1652 the Irish countryside was devastated. Famine and plague were widespread. The 'New Model Army', as Cromwell's soldiers were called, had adopted a scorched earth policy, attacking the agricultural infrastructure and killing livestock. This was met by guerilla attacks from the Tories. In response, retribution was meted out to the civilian population in the form of terrible atrocities. History informs us that the wolf population in the country had thrived on the slaughter and devastation, so much so that a generous bounty was placed on the head of a wolf. This was followed by a similar bounty being placed on the head of a priest. Cromwell's government believed that by getting rid of priests the Catholic religion would die out. Priests went into hiding and ministered to their flock incognito, using 'Mass rocks' in remote areas for Sunday worship. Because so many men had died in the war and so many soldiers had joined the armies of France and Spain, large numbers of women and children were left defenseless and vulnerable. These were rounded up and sold as slaves to work in the sugar plantations of the West Indies. By the end of the war, estimates suggest that over 100,000 men, women and children were captured for sale as slaves to labour in England's expanding empire. Their descendants on the islands of Montserrat, St. Kitts, and Barbados have retained their Irish names and are proud of their Irish heritage. It is estimated that at one time seventy percent of the total population of Montserrat were Irish slaves.

Henry Ponsonby must have been relieved to hear that most of the potential troublemakers were shipped off to the Caribbean, so that his ill-gotten gains could be made secure with imported loyal English peasants. English opinion at the time was quite proud of these accomplishments as can be noted in Prendergast's, *Thurloe's State Papers* (published in London in 1742), "It was a measure beneficial to Ireland, which was thus relieved of a population that might trouble the planters; it was a benefit to the people removed, who might thus be made English and Christian, a great benefit to the West Indies sugar planters, who desired men and boys for their bondsmen, and the women and Irish girls to solace them". Such was the fate of Crotta's peasants; dispossessed and broken, rounded up and shipped abroad like cattle. With their lives torn apart, defenceless and vulnerable; we cannot begin to imagine their suffering and distress.

The country was now ready to be re-planted with Englishmen loyal to the Parliament, but first the land had to be surveyed and mapped. An army doctor named Sir William Petty and a group of surveyors, guarded by armed soldiers, completed a survey of the entire country in a little over a year. Cromwell's government was now ready to go ahead with Plantation. The Plantation resulted in a massive transfer of land ownership from old Irish and Anglo-Irish families to a new Ascendency of English Protestants. And so the old order changed in north Kerry: Cromwell's campaign brought an end to the Fitzmaurice dynasty of Lixnaw, who had been Barons of Kerry since the Norman conquest in the 13th century. The Anglo-Norman dynasty of deStack, who had built a castle at Crotta, would be replaced by a Ponsonby dynasty. Their Norman-style castle would be demolished and its stones recycled to build a new Elizabethan-style residence that became known as Crotta Great House. The descendants of Henry Ponsonby remained in residence there for almost two centuries. A new Ascendency was established throughout the country that would hold its privileged position until the nineteen twenties.

Two brothers, John and Henry Ponsonby were high ranking officers in the Cromwellian campaign and were both handsomely rewarded for their service. John was given a large estate in Tipperary and Henry at 'Crottoe', as listed in the Down Survey. A census of Ireland, circa 1659 of the barony of Clanmorice, registered the population of Crotta (Crottoe) as twenty-four people, eighteen Irish and six English. The census records only the name of the title holder: Henry Ponsonby Esq. Who the other five were we can only surmise; presumably men-at-arms who were willing to defend the 'new master' against Tory activity. History informs us that the activities of the Tories kept the counties of Tipperary, Cork and Kerry in terror for many decades. Despite this, a Ponsonby foothold was established in Crotta on a large tract of land. The Act of Settlement of 1666 confirmed by patent a grant of '2,100 acres of partially wooded land at Crottoe to Henry Ponsonby Esq'. In later times, much of the wooded land, the great oak woods of Crotta, fell victim to Britain's Industrial Revolution, making a great deal of money for the Ponsonby family. The construction of Crotta House was completed in 1669. It was a construction project of major proportions and given the prevailing conditions it is difficult to imagine how it was accomplished. Plantation policy demanded that the local workforce be supplanted by labourers and craftsmen from England. Many names of English origin in the locality suggest that 'planter families' came and settled and endured.

We read in the history books that Tory activity forced many of the adventurer landlords to return to England, leaving their estates in the hands of overseers. Henry Ponsonby, however, remained and established himself as the new lord and master. He married a lady from Athy, Co. Kildare, named Rose Weldon and so began a dynasty that would endure through seven generations. He died in 1681, some twelve years later. After his death his eldest son, Thomas, inherited the estate. He was married to a lady from Ballygahan, Co. Limerick named Susannah Grice and so with the passage of time the new masters became the accepted 'gentry' with power and influence over the lives of their tenants. A cobbled road was built from the Great House to Garrynagore to provide access for traders and workmen coming daily to the estate, while a large number of female staff 'lived in'. In order for Cromwell's plantation to be successful it was necessary to supplant, not just the land and the people but also the Catholic religion. There was a Catholic church in Kilflynn (Cill Flainn in Gaelic means Flynn's church) dating back to monastic times when a hermit monk came to live by the banks of the Shannow river. This was closed down and the Ponsonby family established a Protestant church nearby, which became the place of worship for future generations of Ponsonbys and other

Protestant families and, at a later stage, the Kitchener family. Close by this church, which is called St. Columba's Church, stands a large dilapidated tomb that carries the inscription 'W.P. 1795', in memory of a later Ponsonby heir, William Ponsonby. Each generation of the family were interred in this tomb.

Thomas Ponsonby and Susannah Grice were succeeded by their only son, Richard, who became a member of parliament for Kinsale and consequently was absent from his estate for much of the time. Nevertheless he ran a thriving cider industry on the estate, employing a large workforce in the orchards and at the cooperage. His name, 'R. Ponsonby 1760', is carved into the cider-press used for crushing the apples. The inscription on this enormous slab of smoothly carved rock is the only identifying mark of the Ponsonby family to be found in Crotta today. The walls of the orchard remain and stand as a monument to Richard Ponsonby. He was married to Arabella Blennerhassett but, having no heir, the estate passed on to his nephew, who was the son of Rose Ponsonby and John Carrique of Glendine near Camp. The Carriques, like the Ponsonbys, were also descended from adventurers who had come on the Cromwellian campaign and had been granted land in west Kerry. The estate at Crotta was willed to their eldest son William Carrique in February, 1762, who "thenceforth assumed the name and arms of Ponsonby of Crotto". William assumed the double-barrel name of Carrique Ponsonby and married a lady from Ballyheigue castle named Margaret Crosbie. In turn he was succeeded by their only son, James Carrique Ponsonby who married a lady from Sligo named Mary O'Hara in 1766. He was succeeded by his eldest son, another William Carrique Ponsonby, who was sent to Eton College for his education and later became a major in the Kerry Militia and served as High Sheriff for the county. He was married to Elizabeth Gunn of Ratoo whose lineage, it was claimed, could be traced to Edward III of England. The Ponsonby dynasty came to an end with their son Thomas Carrique Ponsonby, who was born in 1800 and spent some time in the Royal Navy. On the death of his father he took over the estate at Crotta as landlord and local magistrate. When he became insolvent in the late 1840s the estate and its debts were put into receivership, under the Encumbered Estates Act, to a Samuel Julien from London.

The Ponsonbys and their stories have passed away with only an echo of what once was reaching us across the centuries. Apart from names and dates, historical records on the seven generations are very sketchy. Local folklore tells nothing of what dramas and intrigues were played out within these now-ruined walls, or whether or not they were caring landlords who were concerned about their tenants' welfare. But the last of the line, Thomas Carrique Ponsonby was remembered for his part in promoting faction fighting. In the early part of the nineteenth century faction fighting was a feature of rural life throughout Munster. Despite their strenuous efforts, the Catholic clergy were unable to prevent this from happening because many landlords actively encouraged it. One such landlord was Thomas Carrique Ponsonby who actively promoted a faction fighting group in the Crotta area and organised fights between opposing factions. Evidence recorded at the police investigation into the Ballyeagh incident of 1834 suggests that "Mr Ponsonby, a landlord and a magistrate was actively involved on the side of the Mulvihill faction". Fights usually took place at fairs or other convenient gatherings when drink was in abundance. The shillelagh was the chosen weapon. The establishment of the time tended to ignore faction fighting, believing that it was a safety valve for social stability. They were happy to see the Irish underclass fighting amongst themselves and not directing their anger at the system that kept them downtrodden. Some historians, however, believe that this did not have the desired effect as there were strong links between the organised faction fighters and the Whiteboys. The Whiteboys (Na Buachaillí Bána) were a notorious agrarian secret society who sought to address the oppressive practices of landlords in relation to rack-rents, evictions and the reviled tithes.

As well as paying very high rent to the Ponsonby family, all tenants of Crotta estate were obliged to submit approximately one-tenth of their annual produce to the Church of Ireland clergy in Kilflynn. In the absence of money, the tithes could be paid in the form of pigs, poultry or corn. The payment of tithes for the upkeep of Protestant clergy was one of the Penal Laws imposed on Catholics. It was a particularly odious law that created much ill-feeling. These were hard times for labourers and small farmers and consequently the tithe collectors were often targeted by the Whiteboys. Many atrocities took place and those who were suspected of being involved could expect neither mercy nor justice from the forces of law. A police barracks was located at Crotta cross for the protection of the Great House. Thomas Carrique Ponsonby was the local administrator of the law and as there was no right of appeal, tenants were at his mercy. Those who escaped the hangman's noose would face penal servitude for life in Van Dieman's land. But circumstances were about to change. In the wake of the Great Famine, as Thomas Carrique Ponsonby faced bankruptcy he was forced to put the estate into receivership. He and his family had to vacate their beautiful home. Their privileged position had come to an end. The Ponsonby reign in Crotta was over. In 1852 the house and estate were bought by a retired army officer of an Indian regiment named Colonel Kitchener.

Colonel Henry Horatio Kitchener had made his fortune in India and took advantage of the property depression in Ireland due to the Great Famine. From 1846 to 1849 the population of Ireland had been reduced by fifty percent or more; millions emigrated and millions died of starvation and disease throughout the countryside. Rent from the tenant farmers dried up and many landed gentry were left without a source of income. The Encumbered Estates Act 1849 was put in place to facilitate the disposal of bankrupt estates. New owners took over, evicting tenants who were in arrears and clearing many of the smallholders off their land without compensation. The tenant farmers had no legal rights to the land they worked. It was bought and sold over their heads, without consultation. Any tenants that remained on Crotta estate after the famine were to suffer the same fate at the hands of Colonel Kitchener. The new landlord had very different ideas about farming and smallholders were an obstacle to his plans and had to be removed.

He had at first bought lands at Ballygoghlan in County Limerick and had been given the use of a hunting lodge at Gunsborough, near Listowel. His second son was born there and was named Horatio Herbert. He would become the icon of British military conquests over the coming half century and be revered as the great Lord Kitchener. The family moved to Crotta in 1852, where two other sons and a daughter were born. The new owner's policy was to convert, by fair means or foul, the clusters of small uneconomic smallholdings to large tracts of grazing land. Unlike the previous landlords who derived their income from rented holdings, Colonel Kitchener's ambition was to manage and exploit his large farm. Land that had supported hundreds of families was now to be cleared. Evictions were common and his unpopularity was widespread. Historians inform us that the crime rate increased alarmingly at this time throughout the country, with acts of violence becoming commonplace. But the new regime at Crotta succeeded and as time went by became accepted. The extremely large fields of grazing land that were part of my childhood experiences were the result of his policies of land clearance. He had a keen interest in agricultural science and ran his new estate with military efficiency. He took a personal interest in all aspects of the farm and spent most of his day in the saddle overseeing the work in progress. He improved the breeds of cattle, sheep and pigs and cleared and drained large tracts of what was seen as wasteland. The most unusual story that has filtered down from that time was his insistence that he and all family members should sleep between sheets of paper, believing this to be more hygienic than being in contact with the bedclothes; a belief he may have brought with him from the Indian subcontinent.

Colonel Kitchener had very definite and progressive views on education. The young Herbert Kitchener and his siblings had a series of tutors and governesses in their early years. They studied Mathematics, History, Geography, French and German. For the boys, this was accompanied by some instruction in the theory and practice of farming and later they learned the basic arts of estate management. When they got older they attended a school run by William Raymond, the Protestant vicar of St. Columba's church in Kilflynn. It was believed that the future Earl of Khartoum and Sirdar of Egypt was not particularly interested in his education and was happiest on horseback riding around the fields and lanes of Crotta. The four Kitchener brothers spent their days together remaining aloof from other boys in the neighbourhood. On warm summer evenings they would ride their horses to Banna beach where they learned to swim. A story goes that their full length bathing suits were the subject of much laughter from the local boys who swam naked.

Another story relates a more serious incident. On a day when a group of workmen were digging a drain in one of the fields the young Herbert Kitchener rode up on his pony. He played some practical joke on one of the men who had a short temper and who instinctively lashed out at his tormentor, giving him a bloody nose. Fear and consternation immediately descended on the group; striking the master's son could have serious consequences. It was an impulse of anger the workman instantly regretted. The young Kitchener, however, sensing their worry and realising that he was mainly at fault put their minds at ease; he was not going home 'to tell tales'. As the young master he was required to oversee improvement works being carried out on the estate, not hinder progress. He was also required to take on other responsibilities, one of which was to drive his father's cattle to the fairs at Listowel, some eight miles distant. This meant a two-hour walk in the early hours of the morning with the other drovers. According to a local historian, the manager of the Listowel Arms Hotel would be told to provide the boy with breakfast only when he had succeeded in selling the cattle at the price specified by his father.

In 1863 the Kitchener boys were sent to a boarding school in Switzerland, where they were teased about their Irish accents. So instead of mixing with other students they studied hard and became fluent in French and German.

Their mother, who was of French origin, had passed on her knowledge of the language to them during their childhood years at Crotta. During their second year at boarding school she contracted tuberculosis and as her illness grew worse their father decided to move the whole family to Switzerland, but neither the mountain air nor medical treatment helped her condition. After his wife's death he sold his estate at Crotta and settled in Dinan in northern France where his children visited during their school holidays. The young Herbert Kitchener was to choose a military career that would see him serve the Empire in Egypt, India, and South Africa and become a leading figure in Britain's colonial domination.

After his crushing defeat of the Dervishes at the battle of Omdurman in 1898 Kitchener's military genius was lauded throughout the Empire. With only the loss of twenty-four men the British forces had slaughtered over ten thousand Sudanese troops. Kitchener became the hero of the age and over the years was handsomely rewarded for his military services with titles and money. Evelyn Waugh, the acclaimed English novelist, recounts that as an eleven year old boy at Heath Mount school, he sometimes ran errands for the War Office, where he would loiter about to catch a glimpse of the great military genius of the age. Kitchener never married, nor had any children and in 1911 he bought Canterbury Country House, located in Broome Park, for his retirement, but spent little time there. He was dead within a few years. He is best remembered as the recruiting poster image for World War 1, declaring that 'Your Country Needs You'. The stylized image, with piercing eyes and handlebar moustache, that enticed hundreds of thousands of young men to their deaths in France, portrayed the militaristic autocrat and ruthless warmonger that Kitchener undoubtedly was.

Fifty thousand Irish volunteers died in the First World War and few families were unaffected. The 16th Division composed mainly of Catholics from southern Ireland, lost 4,000 men in the assaults on the villages of Guinchy and Guillemont. After this slaughter John Redmond approached General Kitchener to arrange for Irish chaplains to be sent to France to hear the Irish soldiers' confessions before going into battle. It was rumoured that Kitchener was furious with Redmond, stating that the French clergy could do it. When it was explained to him that the language presented a barrier, he is supposed to have retorted "They do it in Latin, don't they". Apparently he had attended a Mass that morning and was confusing the Latin Confiteor recited at Mass with individual confessions. It is generally believed that Kitchener held the Irish soldiers in low esteem and had little compunction in sacrificing them. Those who volunteered had no idea what lay ahead of them in the trenches of the western front.

The obscenity of the trenches was brought home to our family by Paddy Hanlon from Lixnaw. Daddy had been a close friend of his since boyhood days. With the impetuosity of youth, Paddy had signed up to serve in "the war to end all wars" and "to save brave little Belgium". He described the horror of trench warfare in graphic detail; how they "had lived like rats" and with the rats, in the mud and the dirt. The lice, that inhabited the folds of his sodden, foul-smelling clothes, were a constant torment. When it rained, the wet slushy mud kept his feet wet and cold with no reprieve, causing chronic 'trench foot'. Gas poisoning had affected his lungs and respiratory system. The deafening sound of gunfire and explosion was the background noise of every day and night. Fear, terror and death were his constant companions. Paddy had survived it physically but not mentally. Back home he eased his shell-shocked brain with alcohol whenever he could afford it. People said he was "very fond of the drink" and the parish priest at that time was vehemently opposed to drunkenness. A popular story relates that one evening he found Paddy lying in a ditch as drunk as could be and he preached at him saying: "What if the Lord called you now, Mr. Hanlon?" to which Paddy replied: "I wouldn't be able to go, Father". Paddy had responded to Kitchener's index finger and had wasted his manhood in the hell holes of northern France, only to live out his years in misery and suffering.

To distance himself from his Irish background Kitchener is reputed to have used the much quoted expression, attributed to the Duke of Wellington; "Because a man is born in a stable that does not make him a horse". Nevertheless, he came back to see his childhood haunts in June 1910. During a motor tour of Kerry he visited Ballygoghlan, Gunsborough, Listowel and Crotta. The old house at Crotta was still intact and no doubt held many childhood memories within its deteriorating walls. Nostalgia satisfied, he went on to spend some time in Killarney. A short extract from the Kerryman newspaper, dated 25th June, 1910 states:

> *'Lord Kitchener enjoyed his trip through the Lower, Middle and Upper Lakes yesterday, immensely. He stated it had been forty years since he had been in Killarney last and he sincerely hoped he would be able to visit it oftener in future. He was charmed with the scenery and the weather was ideal. His lordship very kindly added his signature to the distinguished roll already hung up in the lounge of the Royal Victoria Hotel, and by special permission, graciously granted to the proprietor, his lordship was photographed in front of the Royal Victoria Hotel before leaving for Dereen,*

the residence of the Marquis of Landsdowne, en route to Cork via Glengariff. His lordship charmed everybody with whom he came in contact in the Lakeland district by his extreme graciousness and affability, and carried away with him the God-speed of everybody'.

'His extreme graciousness and affability' was replaced by cold-bloodedness in April 1916 when he commissioned General Maxwell to put down the Irish rebellion with the same ruthlessness with which he had crushed the Afrikaner rebellion in South Africa. He did not live to see its outcome, however. Later that year, on his way to meet the Russian government, his ship, HMS Hampshire, struck a German mine off the Orkney Islands. All but few were drowned. Kitchener was dead at sixty-five. A requiem service was celebrated in Westminster Abbey, attended by the king and queen and many dignitaries of state. Messages of sympathy to the British people poured in from around the world, but not from Ireland. The name of Britain's great military leader would not be revered in the land of his birth or in the place of his childhood. There are no memorials to him in Crotta or in Kerry, apart from a small prayer book dedicated to his memory in St. Columba's Church, Kilflynn. No street name or square in town or village marks his presence or points to his fame. Field Marshal Horatio Herbert Kitchener, Sirdar of Egypt, First Earl of Khartoum, Secretary of State for War would go unremembered in the locality where he spent his youth. But for us as children his memory was regularly evoked by the humorous quip of our much loved Auntie Sara: *"Make way for Lord Kitchener"*.

Chapter Two

The Past and the Present

The ruins of the old Great House and its surrounds provided a magical, haunting environment for us as children. Its decaying majesty and grandeur filled us with a sense of wonder and awe as we tried to imagine its past. Some exotic trees and shrubs that had survived the utilitarian onslaught of the farmyard added to its charm. The smooth limestone flagstones outside our backdoor and the ornately carved window sills spoke of former grandeur. The enormous tower block that was once the stairwell of the Great House was still standing, proudly defiant and unyielding to the crowbar and sledgehammer. Its ivy clad outline could be seen from any part of our farm. We called it 'the turf house' because its empty shell was used for storing the six lorry loads of turf we burned each winter. In its heyday this was a magnificent arched entrance to the rear of the house. Into this arch was fitted the curved window frames that surrounded the heavy oak door. Above this door was a high rectangular window and further up, reaching to the eave, was another smaller window. On the inside of the ruined building we could see the sloping lines of the staircase running along both sidewalls and ascending to the top of the building. At ground level the staircase faced the front entrance of the house across the main hall. Those walls held the secrets of two centuries of a life-style that had long since passed; a life-style of elegance and arrogance, of wealth and privilege that was paid for by the sweat and toil of Crotta's peasants.

Today, nothing much remains of the Great House proper except an outline of its front wall at ground level. What remains standing and roofed is a slightly lower edifice that was connected to the Great House with adjoining doors. This building provided residential quarters for the household staff who did the cooking, cleaning and daily maintenance of the Great House. This building became our home. It was a very large house; its many rooms were spacious with high ceilings and big windows. Adjoining this building were the stables and coach house, set at a right-angle to the house. At the end of the coach house building was another residential building that housed footmen and drivers. On looking up at the rafters in the loft overhead the stables one can only marvel at the craftsmanship employed in putting the enormous oak planks in place; especially the intricate structure that carried the roof around the corner. These beams have withstood the test of time, carrying the weight and pressure of a heavy slate roof for over three hundred years.

Visiting my old home now brings a lump to my throat and my mind goes racing back through the years. It has been unoccupied since 1991 and decay is all around, as time and entropy carry out their handiwork, unopposed. The bare windows stare vacantly out at a completely changed landscape, stretching off into the distance. The trees and ditches I knew so well as a boy have been bull dozed into oblivion, creating a green prairie for modern farming. As I enter the house my heart skips a beat. Echoes of the past are everywhere but the present reality assaults my senses. Our once beautiful kitchen has become a storage area and a dumping ground for all sorts of farm and dairy equipment. The vinyl floor is covered with dirt and junk and Mammy's shiny enamel range that once breathed life into the house is blackened with sticky wet soot. Hanging on the chimney breast is a small gilt-framed picture. Behind its dust covered glass I can make out the words, written in stylish calligraphy:

> *Christ is the Head of this house, the unseen guest at every meal, the silent listener to every conversation.*

My gaze moves on. Long strips of wallpaper hang foolishly on the walls and the plaster has fallen from the ceiling leaving large gaping holes. The old Pye television set sits on its shelf and waits to be turned on and the little Sacred Heart lamp flickers on the wall, still burning miraculously after all these years. Daddy's old armchair occupies the same spot beside the fire place and his big black boots are underneath it. The built-in press beside his chair holds a few of his things; little Christmas presents received from grandchildren over many years - a peaked cap, a pair of heavy woollen socks, a scarf, a pair of slippers still in the plastic wrapper. He would fondly grasp the little hands that held out these gifts and then store them in his press for a rainy day that never came. My reverie is interrupted by the plaintive lowing of calves. Down in our beautiful parlour twelve calves enjoy four-star luxury accommodation. Something inside me rebels against this destruction of my memories.

The parlour had been our good room and was only used on special occasions. It was a large room, tastefully decorated with a deep russet-coloured wall paper and a matching linoleum covered floor. The ceiling was also papered with a white embossed paper that came down to the dado rail that ran all the way around about eighteen inches down from the ceiling. Five large pictures were suspended by long cords from this rail. An old style three-piece sitting room suite filled one end of the room and a black dining room table with a matching sideboard filled the other. Our parlour was a beautiful room. I thought it might have been spared; that it might have survived as a frozen time capsule of what it once was, but utilitarian farming needs proved stronger.

The calves are cordoned off in wooden cubicles along by the wall on the linoleum floor. The colourful wall paper is mostly intact and the six large pictures still adorn the walls. The bookcase by the fireplace offers the calves a selection of reading material. A series of books by Oliver Strange, about Sudden the Outlaw, sit neatly on a shelf. I flick through the pages and I'm transported back in time. I must have read those books a dozen times. Sudden, the fastest gun in the West, was a long-forgotten hero who had influenced my young life. The light fixture in the ceiling holds the same ornate glass shade since John Joe Quilter wired up our house for the electricity in the mid-nineteen fifties. From their plush surroundings the Friesian calves are calling out for attention. Some poke out their heads through the iron bars. I rub their curly black and white faces and they try to suck my fingers. They are such adorable little creatures that I forgive them for messing up our beautiful parlour.

Upstairs in Sara's room her wardrobe holds a few coats and a hat and on her dressing table is the large luminous rosary beads she brought from Lourdes so many years ago. The drawer underneath is filled with odds and ends accumulated over her lifetime; the bits and pieces that marked her journey through life. My heartbeat quickens and in my head I can hear Sara's soothing voice and it conjures up images and memories of things past. When we were small we shared this room with her. Her presence is palpable and signs of her are everywhere. Her cheap metal-framed bed takes up the same space behind the door. I sit on its musty mattress and say, "Where are you, Sara? Just saying Hello to you". She died peacefully in this bed at ninety three, having held her independence right up to the end. I wander into the next room. This was their room, our parents' room, that for some unquestioned reason we never entered long ago. There stands the bed where our embryonic lives began and where her four eldest were born. In the corner there is an old chest full of our school books and other scraps of paper that marked our little achievements along the way.

16 Chapter Two

The Past and the Present

Chapter Two

I flick through an Irish textbook, Cúrsaí an Lae by Máiréad Ní Gráda, referred to as the Curse of the Day by my brother Pat, for the drudgery it imposed on him. The book is wearing an oilcloth cover, neatly stitched in place by a dedicated mother who wanted the best for her children. She had attached a thumb pad with a piece of string from the top of the book in case a sweaty thumb would smudge the pages. The book falls open where the thumb pad divides and I start to read. The first paragraph is familiar; it is suddenly unleashed from my memory bank of rote learning. I remember having used it in many an Irish composition in later years; the 'flowery paragraph' that I would sneak in amidst my feeble attempts at composition. I rummage through the books. We're all in there: myself, Kevin, Tom, Dan, Diarmuid and Pat, with our names signed on the inside covers in the old Gaelic script in our best childish scrawl.

I dropped the book back in the box and gazed around at the holy pictures on the walls, the crucified Christ, Mother of Perpetual Succour, St. Anne-de-Beaupre, Padre Pio and St Theresa, the Little Flower. These were Mammy's favourite advocates and they were all targeted from time to time, even brow-beaten in times of trouble. She had a great devotion to Our Lady. Her trump card was the *Memorare* in which there was a hint of coercion mixed with ingratiating supplication. As I poked around the room I could see their eyes staring at me. A shudder went down my spine. Unwittingly, my mind wandered back to the nineteen fifties when this old house was like a beehive. Daddy and Mammy had eight children; six boys and two girls. Margaret, the first-born, suffered brain damage at birth and because of this disappeared out of our lives at a young age. She lived out her short life in the care of the Daughters of Charity at St. Vincent's Home in Dublin. The second little girl, named Mary, died within a few months of being born, but six boys grew healthy and strong. Kevin was the eldest, I was second and Tom third. Mammy used to refer to us as the 'first clutch'. After a break of a number of years, because of the death of the little girl, Dan, Diarmuid and Pat arrived in turn; the 'second clutch'. Six boys growing up together created their own dynamic and the word boredom was not in our vocabulary.

At its peak our full household was made up of twelve people: Mammy and Daddy, Uncle Mick and aunties Sara and Lizzie and seven children. I was a year and a few months younger than Kevin and people found it hard to distinguish one from the other, but Kevin's name seemed to have stuck in the public mind, so I was often referred to as Kevin. At first, when neighbours called me Kevin, I would point out that I was John, but as time went on I knew there was no point and I gave up, it wasn't worth the hassle. With such a crowded household there was never a dull moment during the busy working days or the long dark winter evenings as we sat around the blazing turf fire. Before Mammy got in the new Stanley range the fire consisted of a huge log of wood on the concrete hearth, around which the sods of turf were piled high. The smoke ascended through the enormous open chimney. By day you could stand underneath the huge overhang of the chimney and see a patch of blue sky above and often during heavy rain, drops of sooty water would drip on the hearth or sizzle for a while on the hot cinders. On one side of the fire there was room for two of our *sugán* (straw rope) chairs to fit into the alcove and on the other side there was a long wooden chest, with a rug on top, on which Daddy would stretch out, using it as a chaise longue. Over the fire hung a big black kettle on an even blacker crane with lots of black pothooks. The continuous singing of the kettle added a hypnotic tranquillity to the scene. A semi-circle of rough sugán chairs in front of the fire completed the cosy set on winter nights.

In the absence of television and radio we made our own amusement. Recalling events from the past was always popular. As very young children, this often took on an element of competition between Kevin and myself in trying to remember things that would gain adults' attention. Kevin would tell the story of 'the ducklings that got drowned in the pond' and this usually got everybody's attention. Ducks drowning was a bit unusual but it happened. It was common practice to have a hen hatching out a nest of eggs in a butter-box filled with straw. Sometimes a hen would hatch out duck eggs instead of hen eggs and it was funny to see the frustration it caused the poor mother hen when her clutch of ducklings didn't behave as she expected them to behave. One evening as she marched her brood to the far end of the farmyard where there was a pond of rainwater, all the ducklings jumped in for a swim and because the edges were steep and slippery they were unable to climb back out. Mother Hen clucked and clucked for many long hours but nobody came to save her babies. Later, Kevin found the six little golden bodies floating in the water. It was a sad sight; their little lives were over before they had really begun. Mother Hen had lost her family, we had lost the enjoyment of seeing them waddling around the yard and Mammy had lost a valuable source of food.

My earliest memory was of a more painful nature and happened while Kevin and I were playing in the back stall. We had two cow sheds that we called the front stall and the back stall. The back stall had been the coach house of the old Great House but was now converted into a house for cows. The massive stone walls were roughly finished and had jagged lumps of stone sticking out. A timber construction ran from wall to wall. This consisted of five upright poles placed at equal intervals, holding up a long heavy beam and each upright had an adjustable post beside it which, when opened wide, allowed the cow's head to go through. This was then secured for the night ensuring that each cow stayed in her own place. In the manger in front of each head there would be a pile of hay and a quantity of pulped turnips to eat during the night. The hay and turnips were now gone and Uncle Mick had let the cows out and had started on the daily clean-out of dung and straw.

Kevin and I, he's five and I'm four, are racing each other along the manger area. The game is to tip the far wall and return. We're having great fun until I trip and crash headlong against a protruding stone in the wall. I'm lying on the dirty floor, roaring in pain and blood is streaming from my forehead. Uncle Mick picks me up and runs to the kitchen. There's nobody home except Aunt Lizzie and she's still in her bedroom. I'm aware of Mick shouting for her as he tries to wash away the blood and stop the flow. He's pressing hard on my forehead and I'm sobbing and shaking as he sits me on his lap with the wet rag covering my face. Lizzie, a retired nurse, arrives with bandages and takes control of the situation. She cleans and dresses the wound with professional efficiency, but during all this time no words are spoken between them. My open wound has brought them together but some other open wound from the past keeps them apart. Silently, she wraps a white muslin cloth bandage around my head and takes me up in her arms, soothing me with a comforting voice. I'm still bawling so she walks around the kitchen trying to distract my attention with this and that. She takes me to the mirror and shows me my new headgear. I look like one of the Somme soldiers I had seen in the 'Kitchener' book that Mammy kept in the drawer. When Mammy gets home further consternation ensues. She has to see the wound for herself and I have to endure a second round of dressing and bandaging. Lizzie takes me upstairs to her bedroom. She reads a story for me and I fall asleep in her bed.

The wound healed in time but left its mark – a white scar over a little raised ridge at the hairline. During my adolescent years when I was becoming body-conscious I used to point out to Mammy that I could feel a little hump on my forehead. She would brush my concerns aside saying that it was probably a butt of a horn that was beginning to grow there and she would have it burned off with caustic, as she did to the young calves. I was not amused by her attempted levity.

As a child I must have been accident prone. Many of my memories are injury related. On a wet day in early spring Auntie Sara is minding us in the kitchen. The kitchen is big, about twenty feet square, with the huge open fireplace up front. Two large windows provide the light and two heavy oak doors, one to the back and one to the front provide access. These are the original doors, dating back hundreds of years and are still as solid as ever. To keep us amused Sara made up a game: The Wolf and the Sheep. She is sitting in front of the open fire, where the sods of turf are piled high against the burning log. Tom and I take up our positions as 'sheep' at the door and Kevin is the 'wolf' hiding by the window. Sara calls out; "Sheep, sheep come home" and Tom and I answer: "No we can't, the wolf is out" and Sara calls again and we dash across the kitchen floor into her open arms to safety. But on one occasion Kevin is rather too eager as a wolf and comes at me with a rugby tackle and pushes me headlong into the fire. A cloud of ash and sparks erupt and puts a layer of dust all over the kitchen. Luckily a coating of fresh turf had been put on the fire and it was only smouldering rather than flaming. I escaped with only a blistered scalp and singed hair, but poor Sara got a terrible fright. Our fun was over for that day and she never played that particular game again.

In winter the house was impossible to heat. Even though the walls were enormously thick there were many exits for heat to escape and the ceilings were ten feet high. The main source of heat came from the open fire in the kitchen but occasionally fires would be lit in the bedrooms. All the upstairs rooms had beautifully ornate cast-iron fireplaces from times gone by, but the grates were small and could only fit small sods of turf. So the small black turf was used because it gave off great heat. One such glowing fire stands out in my mind from a very young age. I must have had flu or something because I can see myself tucked up in an armchair in the back bedroom in front of the fire, with Lizzie reading a story. Lizzie had worked all her life as a nurse in Greenlawn Hospital, Listowel. She had never married and was now part of our extended family. Nursing had been her life and now in her retirement she seemed to find fulfilment in caring for us when we got sick. Lizzie was a tall, slender, slightly austere lady who, for much of the time, remained aloof from the goings-on in the house. She had her own bedroom upstairs at the end of the corridor where she spent a good deal of her time reading.

Instinctively, I know she's happy to be looking after me, away from the hustle and bustle of the kitchen and I'm relishing the care and attention that's being lavished on me. She's toasting the bread in front of the fire using a long-handled toasting fork. Having buttered it generously she cuts it into small squares and hands them to me one by one. They're delicious and I wash them down with a mug of hot sweet tea. My child's voice is saying, "Make more toast, Lizzie" and I ask her to let me hold the toasting fork to the fire. I pretend to be sick for many days just for the sheer luxury of having Lizzie pamper me. At regular intervals she gives me fresh orange juice to drink and a spoonful of deliciously sweet glucose to build up my strength. She cuts the oranges in two and presses each half down hard onto the little glass dome of the juicer until all the juice is squeezed out. The sweet aroma of oranges fills the room as she throws the skin into the fire. It's time for another story from her big book of children's stories that has fascinating illustrations on every page. As well as the wonderful stories I'm hearing I'm also doing a little artwork. Lizzie is good at art and draws pictures of birds and animals, with mountains and woodlands in the background, for me to colour in.

As I'm getting better I spend some time looking out the back window. The window is big, about four feet by six feet, with large ornate casements on either side. Lizzie lifts me up on the wide window sill and from there I have a good view of what is happening outside. I see a man picking stones with his bare hands in the back-lawn field. The field is littered with stones and he is piling them into little heaps. The man's back is stooped as he works but every ten minutes or so he straightens up and swings his arms to make them warm. He rubs his palms together for a little while before continuing his work. I feel sorry for him, being all alone in this very big field, doing what seems like an impossible task, with only the crows for company. Even though the room is very warm I feel the cold air coming off the window panes and I'm happy to be inside.

At the far end of the 'haggard' I see Sara washing a bucketful of spuds for the dinner. She doesn't put her hands in the cold water, but instead is using the handle of a shovel, stirring the spuds around in the bucket to get them clean for boiling in the big black pot. Far off in the distance there are cattle waiting at a gate. They are waiting for Mick to bring them fodder. I watch Mick at the hayshed as he places a long piece of rope, looped into two strands, on the ground. He makes up a very large bundle of hay. Then he squeezes the rope tightly, slings the bundle over his shoulder and walks back through the field. His black wellingtons are splashing water as he wades through the puddles. The wind is whipping at his coat as he struggles along with his load. The cattle are hungry and are bellowing loudly. He trudges his way into their field and deposits the hay on a patch of dry ground. The cattle stop bellowing and Mick returns with the rope slung over his shoulder. These scenes get etched in my memory as Lizzie lifts me off the window to listen to yet another exciting story by the fireside.

Some stories I listen to again and again, even though Lizzie is tired of reading them. They take me to strange faraway places, to countries whose geographic locations are beyond my understanding. Androcles and the Lion is a story from ancient Greece in which a runaway slave makes friends with a lion. They share a life together for many years but later the slave is caught and sentenced to be torn to pieces by a wild beast. Even though I know the outcome of this story I'm listening with nervous apprehension as the door of the lion's cage is opened. Lizzie's voice is building to a crescendo as the hungry lion is leaping out. Luckily the lion remembers his old friend and saves me from collapsing on the floor with a minor heart attack. My second favourite story brings tears to my eyes every time she reads it. Hassan and his magnificent white horse are captured by slave traders and taken far away across the desert. That night the man is bound hand and foot and the horse is tethered to a pole. When everyone is sleeping the man whispers to his horse and it breaks free of its tether. Then gripping the man's clothes with its mouth the faithful animal drags him all through the night, across miles and miles of sand, until they reach the safety of his home. I beg Lizzie for one last story. She reads about Ferdinand the Bull who prefers to smell the flowers rather than fight in the bullring. As this story ends my eyes are closing; I'm ready for bed. Tomorrow will be another memorable day. I'm so contented in this safe, cosy environment that I want it to go on and on.

Chapter Three

The Orchards

For us as children the most direct link to Crotta's past glory was the abundance of apples at our disposal. The forest of apple trees, guarded by ten foot high walls, told a story of other days. Generations of Ponsonbys had developed a thriving cider industry and planted a wide variety of apple trees. The vast expanse of orchard provided a valuable source of income for the estate. The Great House and its way of life provided most of the employment in the locality. It was the hub around which the local community revolved. A large work force was employed in maintaining the trees and in running the industry. As was the custom at that time, part of their wages was paid in cider. Throughout the winter a key part of their work was pruning to contain the growth of the trees and to achieve the right balance of growth to fruiting potential. A large one-roomed stone building with an open fireplace was home to the caretaker and his family. The apple trees were spaced about fifteen feet apart which allowed cattle graze amongst them to keep the grass in check. The autumn was the busiest time. Cooking apples and dessert apples were kept in a storage house at the side of the small orchard. The cider apples were brought to the cider press. Here the apples were washed, sorted and pressed and the juice collected for fermentation into cider. The cider was then stored and matured in large wooden vats which were housed in the cider sheds on either side of the large arched doorway at the main entrance to the orchard.

As children we used to play on the cider-press, this large, smooth stone slab, pretending it was a stage on which to recite poems and songs, or roll marbles along the apple juice channel. In the 'big orchard' we had another play area, which we called 'Danny Boy's house'. This was the remains of a large glasshouse and because it had a number of heavy iron water pipes strewn around it we christened it 'Danny Boy's House' and we used to sing "T*he Pipes! the Pipes! are calling"* whenever we were near it. Back then the small orchard still had most of its trees while only a scattering of trees was left in the big orchard. We were totally spoiled for choice with

Chapter Three

the abundance of apples at our disposal and we wastefully indulged our taste buds, choosing from this or that tree and then throwing the apple away with only a bite taken. There was a variety of cider apples, cooking apples and dessert apples but one particular variety stood out above all the others. These grew on three large trees halfway down the middle row; they were very large yellow apples almost as big as the Brambley cooking apples that grew by the far wall. They were exquisitely juicy and sweet and we gorged ourselves on them all through late summer and autumn. Occasionally, an apple battle would break out between us using the windfall apples as missiles. We would take cover behind the tree trunks and pelt the apples at each other till our stash of ammunition ran out. As well as apple trees there were pear trees, plum trees and cherry trees. We had little or no appreciation of the treasure trove handed down to us from those far off days.

Some years, Daddy used to sell most of the produce of the orchards to Mr. Brandon who was a fruit merchant in Tralee. He would come with his helpers and strip the trees bare, but there was one tree they were forbidden to touch. It was a special tree whose apples did not rot, but lasted all through the winter, so that we would have apples for lunch at school. When I was in second class, Mrs. Rice would bring a bag of windfall apples from her garden as an incentive for us to do our best. If you got ten arithmetic questions correct you got an apple. One day all my classmates were going up to her desk with their copybooks and coming back down with their prize. But I was stuck on question eight and time was now running out and panic had set in; my mind had gone blank. Copying was strictly forbidden and punishable, but in desperation I whispered to a third class boy, Brendan Twomey, if he could help me. For Brendan it was a 'piece of cake'. He gave me the answers to questions eight, nine, and ten in an instant. I was the last up to her desk but that didn't matter; I was overjoyed to be able to claim this one slightly battered apple, even though I had far better ones in my schoolbag.

In the summer of 1955 Miss Ponsonby came on a visit to Crotta to see her ancestral home. She came from Whitehaven in Cumbria with a younger woman who was her travelling companion. We were playing hurling in a corner of the field by the orchard wall when Daddy came through the gate accompanied by the two foreign ladies. They were both dressed in fine clothes and looked very elegant. Daddy regarded it as his duty to act as a tour guide whenever anyone came to see the old Great House. They walked along by the orchard walls which run the full length of the 'back lawn' field. A certain part of the wall seemed to hold their attention. These were red clay bricks fitted into the stonework with small square openings about three feet apart. Daddy explained that a tunnel ran through the centre of the wall for thirty feet or more. They were deep in conversation as they viewed this strange feature and their tour guide didn't know what purpose the little tunnel served. For many years these red bricked pigeon holes had guarded their secret well; nobody could offer a convincing explanation of why they were there. They were in fact an ingenious method of heating the wall by lighting a fire inside in the tunnel. This would protect the soft vines on the other side of the wall from late spring frost that might damage the fresh green shoots.

As they retrace their steps back to the gate, they stop to watch us playing hurling. Aware of being observed we try to show off our hurling skills until we lose the ball in a forest of briars and nettles growing nearby. While we are searching for the ball, Miss Ponsonby asks in a 'strange' accent what the game is called and in my broad Crotta accent I say 'hurlan'. She asks for my hurley and swings it gently against the grass as if she were striking a ball. She gives it back, she smiles, says goodbye and they move on. Daddy escorts them through the main arched entrance into the orchard and I follow along behind listening to their conversation. They stop to look at the cider-press; the massive lump of flat rock with the channel running all around the edge towards the tapered end. Chiseled onto the side of this stone is the inscription: R. Ponsonby, 1760. Miss Ponsonby is delighted to have made the connection. She talks excitedly and takes many photographs of the inscription on her little box camera. They walk to the end of the small orchard and peer through the gate into the big orchard. She asks some questions about the few remaining trees and the large pond further down.

Then Daddy takes them to the 'well field' to see another circular pond, with a small island in the centre. It's overgrown with briars and bushes but from one spot they can see the outline of the island and hear the gurgling flow of the water. Daddy explains that it had been a recreation area in days gone by, where members of the Great House could while away warm summer days, boating in the circular river. Miss Ponsonby looks at her watch. It's time to go. They don't want to keep the hackney car waiting too long. She thanks Daddy for his time and courtesy. He shakes her gloved hand and says "Be sure to come again". The driver seems impatient. He starts the engine as they are about to sit in. As the car moves off I see a white gloved hand waving from the window and I wave back.

Daddy's description of the boating had filled up my head with images of

Ponsonby ladies gazing leisurely at the water as their boat drifted around the little island. So the following day I decided to try my hand at boating. In the big orchard there was a large pond where cattle came to drink. The pond was broad but not deep, only two feet in the deepest parts and the water was perfectly still. I didn't have a boat so I had to improvise. There was an oval-shaped galvanised bath tub hanging in the shed. It had been used for bathing us when we were small. It was three feet long and two feet wide with sloping sides. This would be my boat and an old sweeping brush, with a short handle, would be my paddle.

Off I sneak to the pond when no one is about. I place it on the muddy edge and gingerly get in. By pushing the brush handle against the bank I ease it gently into the water. It's a bit wobbly and needs very careful handling but I get to the centre of the pond and sit there looking at the ripples in the water. Tall rushes grow on one side where water hens make their nests in the spring. The other side is covered with long green pointed plants that we use for dueling and sword fighting whenever we pass by. The sun is shining, the scene is peaceful and it's lulling me into a daydream. Suddenly my reverie is interrupted by a noise that instinctively sends a shiver down my spine. It's a moving sound from somewhere at the back of the pond that I recognise immediately without looking around. The bull has come to drink. I hadn't realised he was in the orchard and now it's too late. My heart begins to race.

He stands there staring at me, swishing his tail to keep the flies away and snorting air through his nose. Eventually he reaches down his head to drink and when finished starts to graze around the edge of the pond. I seem to be of little interest to him. I inch my boat a little further away and sit there motionless for what seems like hours, willing him to walk away to the far end of the orchard. He's quite contented to stay and I know I'm trapped. My mind is racing and I can hear my heart pounding. My whole body aches but I dare not shift my position in case my flat-bottomed boat overturns. Just when desperation is beginning to take hold, Pudzie appears out of nowhere. Pudzie, a mongrel fox terrier, has enough of the terrier gene in him to be aggressive and fearless, regardless of the size of the enemy. Instinctively, he seems to know what to do. He runs at the bull snapping at his heels and growling. Anxiously, I watch this David and Goliath duel. The bull turns to charge in an attempt to intimidate him but brave little Pudzie puts renewed energy into his guerrilla attacks. His teeth pinch the bull's hind leg and he lets fly with a kick that sends Pudzie into the air. I gaze in horror as I hear him yelp, but he is prepared to go down fighting. He redoubles his efforts, running from side to side to confuse the bull who eventually accepts defeat and runs off to the far side of the orchard. I scramble ashore in drenched shoes and socks leaving my boat capsized in the water. I pick Pudzie up in my arms and squeeze him so tight that he lets out a little yelp of pain. His little bruised body is sore and there's blood at the edge of his mouth. He would need careful nursing for a few days. When I get home I place him in a box by the fire. I'm afraid to tell anybody what really happened and so Pudzie and I keep our secret for evermore.

Later I retrieved the bath tub and placed it by the side wall of our back-kitchen where it would collect the raindrops from the tin roof. It would serve a different purpose now. Using a large jam jar I caught minnows in the stream and put them into it. I set about creating a miniature of the circular pond by making an island of stones at the centre. I added a few sprigs of watercress and other water plants around the edge to make it more natural looking. Then shaping a cigarette packet into a tiny boat I made up a figurine of Miss Ponsonby and put her floating in the water. I watched her sailing round and round with the little shoal of minnows racing ahead at great speed. I sat there lost in childish daydreams. I thought she would come back to Crotta but she never did. The dilapidated ruins of her ancestral home were not worth a second visit. But her memory lived on in my head and many years later, while attempting to research the history of the estate, I sent a vaguely addressed letter to Cumbria and was greatly surprised to get a reply. It came from a solicitor's office. It was brief and to the point; Miss Ponsonby was not in good health and had no wish to have any further communication.

Chapter Four

Family History

Jeremiah, our father, was the youngest of a family of eight who had grown up on the family farm at Ballyrehan, whose tenancy dated back to the eighteenth century. It had once been part of the 2,100 acre estate granted to Henry Ponsonby. In later times, when extra income was needed Ponsonby landlords sometimes sold off parts of the estate. Somewhere along the line, when the Penal Laws on Catholic ownership of land were relaxed, the Galvin tenants became the owners of their farms. Presumably the farm at Ballyrehan was large enough and diversified enough for its occupants to survive the ravages of the Great Famine. Daddy's father was the third generation of Galvins on the farm. His name was Michael but he was known to everyone as the 'Saggart' (the Gaelic word for priest) because he had spent some years studying in Maynooth. For reasons that are long since lost he gave up his ecclesiastic studies to work the family farm. He married Maggie O'Shea from Moyvane and they had eight children: Johnny, Mick, Lizzie, Sara, Pidge, Jeremiah, Sister Kevin, and Sister Jerome.

These two daughters had joined religious orders. It was the Church's practice to give the names of male saints to nuns, so we always referred to them by these names. Sister Kevin was a member of the Mercy Order in Killarney. We never had a chance to know her but from Sara's stories we gathered that she was light-hearted and funny and didn't take herself too seriously. She died shortly after my brother Kevin was born and hence the name connection. He was christened Michael, after his paternal grandfather, with Kevin as a middle name, but Sara requested that he be known as Kevin in memory of her much loved sister. This was a break with tradition, as all our names followed traditional lines, giving priority to the paternal grandparents, then the maternal grandparents and so on down the line.

Sister Jerome had joined the Holy Faith Order in Dublin, which was a teaching order, and lived out her life in their convent at Clarendon Street. But she often came on a one-day trip to visit us. Her nun's habit was dark brown in colour with a very large hood. The hood projected about twelve inches forward on either side of her face so that she had no side vision; the purpose of which we did not understand. Daddy compared it to the winkers on the horse that obscured the horse's vision from distractions. As small children we thought it funny to be so covered up and sometimes played tricks on her from behind her back, but she was patient and gentle and enjoyed our silly games. Mammy was fond of her and occasionally visited her in Dublin. Some weeks before I went to secondary school she sent a little parcel containing a prayer book with my name inscribed in beautiful calligraphy. It never got much use but through the years some strange sense of duty or loyalty protected it from being dumped. Many years later during my student days in Dublin I got to know her fairly well. At Sara's request I used to visit the convent when time permitted. She would have a big meal ready for me and would summon some other nuns to the reception room to meet her nephew and watch him eat. They would sit around plying me with questions as I worked my way through a big main course and an even bigger dessert. They seemed so interested in my life that I often jazzed up my stories just to make them laugh. When leaving, she would keep shaking my hand for longer than was comfortable, while looking into my eyes. She lived well into her nineties and is buried in the Holy Faith plot in Glasnevin cemetery.

Johnny was the eldest of the three sons and following tradition, inherited the family farm. He married late in life and had no children. Mick was the second eldest son and never married. Lizzie, the eldest in the family, also remained single. She worked as a nurse in Listowel for most of her life. Pidge died at a young age and her story was lost to us. Sara was married but was tragically widowed in her forties. Jeremiah, the youngest, was tall and straight and lean and, at one time, very athletic. As a young man he was very interested in 'coursing' and greyhound breeding and trained many greyhounds with a modicum of success. But his early life was mostly a closed book; we only got a glimpse into his past through whatever bits of information Mammy gleaned from friends and colleagues. We know that he was given the opportunity of secondary education at St. Michael's College, Listowel, where he attended for three years but rumour had it that he spent more of his time around the town than he did in the college. He spent some years working on a neighbouring farm, which belonged to Maurice Galvin, his first cousin, who had an all-female family and needed male help with the heavy work. Later, he worked with his older brother, Johnny, on the family farm before the farm at Crotta, through a strange and tragic twist of fate, opened up new opportunities for him and for generations of Galvins to come.

Sara's tragedies created these opportunities. She was married in 1926 and widowed in 1930 and had buried two still-born babies. Her husband, Terence Carmody, along with his brother Batt had bought the remaining lands attached to the Great House after the Land Commission had subdivided much of the estate. The brothers divided the hundred and thirty acres between them and also the remaining parts of the buildings. Making money from farming at that time was difficult so to subsidise farm income, Terence Carmody began demolishing parts of the Great House, to be sold as stones and rubble for road repair or house building. According to local lore, the house had been unoccupied for a number of years and had fallen into disrepair. Because it was remote and deserted it was used as an overnight shelter for freedom fighters on the run during the Irish War of Independence (1919-1920). Consequently, British auxiliary forces set it on fire. The roof was damaged beyond repair and so began its destruction.

A story existed that somebody would be killed in the demolition of Crotta House and, like many self-fulfilling prophecies, this came to pass. Terence Carmody was killed by a huge chunk of falling wall, crushed to death in front of Sara's eyes. Devastated, grief-stricken and alone she looked to her own family for comfort and assistance. Mortgage payments had to be met and so it was essential that the farm would continue to operate. Someone had to take responsibility for the planting and harvesting of crops and the husbandry of animals. Jeremiah, her youngest brother, came to her aid and for twelve years ran the farm at her behest. As Sara was getting older she signed over the farm to him. He took on the mortgage repayments and built a new house for herself and Lizzie in the Barrack field, near to where the RIC barracks had been located. Later Mick left the family homestead at Ballyrehan, where he had been working on the farm with his older brother, Johnny, and threw in his lot with his younger brother at Crotta. Jeremiah was now an eligible bachelor in his mid-forties when Madge, my mother, arrived on the scene. She was in her late twenties and had come on an extended visit to her cousin Mary Whelan, who lived on the neighbouring farm at Ballyrehan. The husband, Gerry Whelan, proved skilful at matchmaking and he brought them together. They got married in 1943 and in so doing saved the Galvin line from extinction. The future of the Galvin gene rested on Jeremiah's shoulders. No other member of the family was in a position to produce offspring. But Jeremiah and Madge 'saved the day' and kept the bloodline going by producing eight children; six of whom survived.

Sara and Lizzie were living in the newly built house at the end of the Barrack field. They were both in their senior years but were strong and healthy and had been used to active lives so they spent most of their days up at the old house helping Daddy and Mammy with the work. On summer evenings they would walk back to their own house through the fields and we as children would accompany them. The dirt track went through a patch of scrubland covered with briars and hazelnut bushes before it reached a rushing stream bridged by two very unsteady planks that Daddy had put there as a footbridge. In the late autumn they would pick the ripe hazelnuts for us, which grew in abundance on one side of the stream. On the other side there was a forest of briars covered with big sweet blackberries, which were easy pickings. When we got to the house we feasted on the hazelnuts, which we cracked open on their kitchen floor with a small hammer. They never minded the mess we made. They sat by the table and enjoyed our antics as we filled their house with noise and laughter. We were the children they never had; they loved us, indulged us and spoiled us.

Lizzie had worked all her life at Greenmount Hospital in Listowel. After retiring she was often hired to do private nursing for local people. One such employment took her to an isolated farmhouse on the side of Stacks Mountain, where an elderly, bed-ridden widow needed twenty-four hour care. Her only son lived some miles away with his wife and young family, so Lizzie was hired to do the night shift. It meant sitting in the old lady's room through the night and assisting her with toilet problems and so on. Lizzie had visited the house the day before taking up her duty and had spent some time getting to know her patient. The old lady told her that she was very lonely because Jack, her husband, had died recently after a lifetime together. On the first night the son stayed in the house to keep Lizzie company and to show her how to work the oil lamp in the kitchen and bedroom. Lizzie assured him she could handle the situation from then on. She was well accustomed to sickness and death from her years of working with geriatric patients and wasn't easily frightened. On the second night Lizzie checked the doors and windows and lit the lamps. She then went upstairs and settled down in the chair beside the old lady's bed. In the shadowy stillness of the room the hours dragged out slowly while the lady in the bed slept fitfully. Lizzie dozed too but awoke with a start at the sound of the front door being opened and heavy footsteps on the stairs. Frozen to the spot she heard the footsteps come right up to the bedroom door and stop. The old lady reached out her hand and grabbed Lizzie's elbow tightly."Don't be frightened, Lizzie", she whispered, "it's only Jack, he often comes back to see if I'm all right".

On sunny summer afternoons Sara and Lizzie would get the pony and trap ready and take us to the seaside at Ballyheigue. They would walk along the sand holding our hands in case anything might happen to us. When we wanted to paddle in the shallow waves Sara would tell us to keep looking up at the houses in the distance in case the movement of the waves might make us dizzy. It probably was her own fear of the water she was expressing but the over-protection and mollycoddling continued throughout our childhood years and it fed into a timidity and nervousness that some of us carried with us through life. The pony and trap was also used for getting to Lixnaw Mass on Sundays. All of us went to Lixnaw church except Mick. The little white pony had a heavy load but would trot steadily all the way. It was so exciting to stand at the front of the trap looking at her flowing mane, feeling the breeze on my face and listening to the sound of her hooves on the road. But the journey home was the best bit because Sara would always buy a big brown-paper bag of 'Kerry Creams' in Jack Mac's. The biscuits would take the edge off everybody's hunger till we got home. In those days people had to fast from midnight to receive Holy Communion in the morning. The Mass was at 8.30 a.m. but for the farming community this meant getting up at six o'clock to have the cows milked in time.

After several years of traipsing back and forth Lizzie and Sara abandoned the house in the Barrack field and moved in with the rest of us. This was not a problem as there were four very large bedrooms. By this time there were four children: Margaret, Kevin, John and Tom. Margaret was the first born; a home-birth with Maggie O'Brien attending as midwife. The labour went on for God knows how long and consequently poor little Margaret suffered a lack of oxygen and got brain damaged. This was a heavy blow, mentally and physically for Mammy but Lizzie helped to lighten the burden. Nursing had been her life and she now had lots of time on her hands. Margaret became her baby and as she grew up she was very attached to Lizzie. She slept in her bed and as she grew older spent the whole day by her side. On wet days she would sit quietly on a chair beside her, rocking herself back and forth. As the years went by they could often be seen walking together along the hedgerows, picking flowers or berries, one getting old and feeble, the other coming into puberty, with the mind of a child. Then out of the blue, the parish priest made arrangements and convinced our parents that it was best to have Margaret placed in a nursing home in Dublin, in the care of the nuns. In those days, the Church was involved in every aspect of people's lives and sexual morality and scandal were top of the agenda. Higher motives were attributed to the priest's intervention and the underlying rationale was never discussed. Mammy was openly distraught but agreed and Daddy went along with her decision. Early one morning, Matt Stack, a loyal and trusted friend, drove them to Dublin to deliver Margaret to her fate.

When they arrive back Mammy's eyes are red from crying and her face is raw and blotched. Daddy sits by the fire in silence, his grim face focused on the flames. He doesn't know what to do or say, or how to comfort his distressed wife. She cries all night long; not a silent cry but a loud wailing cry that fills the house with tension and fear. In between each wail there is a long pause. Waiting for the next wail is nerve-wracking. I cover my head with the pillow to block out the sound, but it goes on and on. Sleep comes slowly for all of us. For many days we exist in a strange subdued mood and each night we anticipate and endure the same fear. And far away a little twelve-year old, brain-damaged girl cries herself to sleep in a lonely room.

A dense cloud of sadness hung over our house for weeks. Mammy cried and cried and Lizzie retreated deep into herself and stayed in her room. Lizzie was never quite the same; her sense of loss weighed heavily upon her but she bore it silently. Sara was the strong one; she kept the house going and kept things ticking over until the healing hand of time soothed our pain. Daddy and Mammy visited the Home in Dublin, regularly at first, but as time passed, less and less frequently. And as the years went by we blotted Margaret out of our lives. We scarcely ever spoke of her. A sense of shame and regret hung over us that we didn't quite understand and couldn't articulate. Those were less enlightened times when a mental disability was something to be kept hidden. Many years later, when I was at college in Dublin, Mammy came to visit her and I accompanied her to the Home on the Navan Road. The nuns had prepared Margaret for our visit. She was neatly dressed and seemed contented but was too scared to come near the two strangers who were looking at her and trying to connect with her. My heart ached and I kept thinking of what might have been and hoped that some kind nun had given her the love her family had not given her. She died a few years later and we brought her back to the family plot in Kiltomey cemetery.

Three women in the one kitchen might not be seen as a harmonious arrangement, but it worked well in our house because they took advantage of the situation. Lizzie and Mammy would go to Listowel almost every Saturday, while Sara would do the cooking and mind the children. Lizzie had worked and lived in Listowel all her life and she needed a regular fix of her old haunts. The pony and trap was the means of transport. As we got a bit older we often had the exquisite delight of travelling with them. The ride in the pony and trap was an exhilarating experience and the clip, clop of the pony's hooves was music to my ears. They would tether the pony by the Protestant church in the square, where the youthful Lord Kitchener had stood guard on his father's cattle until the sale was complete. I rubbed the pony's soft black nostrils and told her to wait there as we headed off for an exciting day of shopping. I loved the old style hardware shops; Carroll's or Carey's for bits and pieces of household equipment and then on to Lynch's for the groceries. Lynch's shop had a very large floor area with heavy wooden counters running along three of the walls, with two assistants behind each counter. At the far corner was Moll Keane who was a special friend and they would spend a long time talking. She sold homemade butter which was especially good. Moll would cut off little bits for me to taste. It was delicious; having that extra saltiness that the creamery butter lacked and it whetted my appetite for the upcoming lunch. My lunch order was simple – crispy, golden chips. At home we had boiled potatoes or mashed potatoes, but never chips. After lunch came the boring bit; visits to drapery shops and having to sit quietly for what seemed like hours listening to conversations about the quality of the cloth. After a quick visit to the Listowel Arms to use the toilet facilities we were ready for the road. The long ride home was made short by the bag of 'Bull's Eyes', or the sticks of 'Peggy's Leg' or the long liquorice 'Blackjacks' that turned my tongue, teeth and lips a shade of dark brown.

Sara, on the other hand, preferred the town of Tralee and would always go on her own. The primary reason for her trips to Tralee was 'to get her hair done'. She was proud of her head of dark hair which never showed any sign of greying. She would catch an early morning bus at Crotta Cross and would arrive back at half past six in the evening. We knew the bus schedule well and made sure to be there to meet her, knowing that she would have a selection of sweets for each of us. Woolworth's sweet counter in Tralee offered a wide variety of sweets, from hard-boiled shiny ones to multi-coloured soft ones. Sara would give each of us our own bag of sweets, so no sharing was required, just some swapping to try out different flavours. Regardless of whether she gave us sweets or not we loved her to bits and she loved us dearly. Never a stern or unkind word emanated from Sara's mouth. She was always positive and joyful in spite of the cruel hand life had dealt her. Whenever misunderstandings and tension arose between herself and Mammy, Sara would let it go over her head, would never retaliate, and would carry on as before. She was an incredibly strong person both physically and mentally who gave freely of her time, her energy and her resources. When we were small we would run to her for comfort with an injured arm or leg or whatever and her welcome would always be, "What's the matter, lanaveen (little child)"? and we would fall asleep on her huge lap, safe and snug and warm.

In later years Sara developed cataracts in both eyes and after the operation her eyes were sore and her sight was poor. Her indomitable spirit was temporarily broken. Mammy would take her to Glendahalan Blessed Well that was reputed to have healing powers for sore eyes and for sight. She would dutifully and patiently bathe her eyes with the clear spring water and pray for healing. Partial vision returned and, in time, Sara adapted and continued to work as hard as ever both inside and outside the house. Bright sunlight always seemed to hurt her eyes so on sunny days she would have to wear dark glasses all the time, even inside the house, and this obscured her vision even more. Nevertheless, this stout-hearted woman carried out a routine of chores without question. One of her daily chores was to empty the milk trough of the remaining sour milk from the previous day and have it clean and ready for the skim milk, which Mick brought back from the creamery. This was a slow and tedious job, scooping out the milk into buckets with a saucepan and taking it to the pigs.

The milk trough itself was an amazing piece of work. It was a remnant from the past, ingeniously carved out of a huge piece of solid limestone. The amount of time, skill and energy put into the making of this artistic piece was hard to imagine. Some Ponsonby landlord had employed a skilled stone mason to seamlessly sculpt the trough on site. It stood by the wall of the stables and measured five feet long, three feet wide and two feet deep. Its walls were about four inches in thickness. When Sara had washed and rinsed the trough the fresh skim milk was poured in and this was used during the day for feeding calves. She would then thoroughly wash and scald the milk churns to ensure there were no bacteria remaining that would turn the fresh milk sour the following day. After dinner she would wash up the ware and cutlery. There was always a lot of washing up to be done in our kitchen but male members, young or old, were never expected to help out. And to our shame we never did.

Of all the chores that Sara took upon herself to do, one was most odious. Prior to the installation of running water and flush toilets, each bedroom had a white enamel urine bucket that had to be emptied daily; else the ammonia smell became unbearable. Sara took responsibility for 'slopping out' from all bedrooms. As she was coming through with a bucket in each hand she would humorously call out, "Make way for Lord Kitchener" so that everybody would give her a wide berth. We, as children, would repeat the phrase back at her, unaware of its origins until many years later. She was a second mother to us and we loved her dearly. Our younger brother Dan probably loved her most. As a small child he slept in her bed and in his baby talk referred to her as 'Dego', a name which stuck for the rest of her life. She died at ninety four, independent and self-reliant to the last. A few weeks before she died, she staggered a little as she attempted to climb the stairs to her bedroom. Diarmuid, by then a young man, was ever watchful of her. He ran to her assistance only to be shrugged off. Defiantly she said "I'm alright, I don't need any help". Her passing left a huge void in our lives for a long time; but her memory lives on and often at family gatherings, things she said and did are re-visited and explored with affection and laughter.

Uncle Mick played a crucial role in our family. He lived with us all of his life and occupied the downstairs bedroom. On Sundays he always went to the 11.00 o'clock Mass in Kilflynn Church, which was about half an hour's walking distance. Afterwards he would go to Parker's for a few pints and arrive home in the afternoon. This was one of the few small pleasures for a bachelor brother of no independent means, who had thrown in his lot with his younger brother as a farm labourer. Mick worked hard and conscientiously for the six days for whatever handout he received at the end of the week. He must have been glad that the Catholic Church had decreed that Sunday should be a day of rest, free of manual labour. Mammy, who loved him dearly, would keep his dinner hot for him. One Sunday, when I was five, I sat on his knee while he was waiting for her to put the dinner on the table. To everybody's amusement, I declared loudly: "There's a strong smell of Sunday from you". They all guffawed and I was chuffed to have made a remark that got everybody's attention. I had disclosed his secret. In those days our house was mostly an alcohol-free zone and we as children were not exposed to such 'evils', except for the one crate of bottled Guinness that arrived at Christmas, along with the lemonade and the raspberry cordial.

Mick was fond of the Guinness but his drinking wasn't usually a serious problem. However, one particular Sunday, he indulged a bit too much and Daddy must have sensed that something was wrong because he set off to look for him in the pony and trap. I asked where he was going and if I could go with him. To my surprise he opened the little door of the trap. I climbed onto the metal step and jumped in. Daddy slapped the pony hard and she took off at a gallop. An open stream ran alongside the road and we were moving at top speed when something frightened the pony and before we knew what happened we were lodged in the middle of the stream.

Luckily the trap had not capsized or fallen sideways. Miraculously, no damage was done and we were both alright, except for the shock. With a few soothing words to the pony, Daddy succeeded in getting the trap back on the road and we continued at a slower pace. About a half-mile further on we found Mick stretched on a grassy patch near Buckley's gate. His Sunday suit was muddied and bloodstained and his fingers were skinned and bleeding. He must have fallen many times on the rough road. He had made it that far but was incapable of going any further.

Not a word was spoken. Daddy managed to get him into the trap and tried to keep him upright as we made our way home. Now and again Mick would mutter something incoherent and would ask me to sit near him, but I was too frightened and upset. When we got home Daddy helped him to his bed. He lay there muttering, still wearing his muddied suit. I had no understanding of what was happening and Daddy offered nothing by way of explanation, either then or later. But that scene by the roadside of my fallen hero disturbed my thoughts for many months. I think I could have taken the truth on board but I was never told and never asked. 'Children should be seen and not heard' was very much the motto in our house.

Whether or not Mick had any romance in his life remained a mystery but Mary Flaherty was somebody very special to him. Mary shared the Flaherty family home with her brother Frank at Crotta Cross, near to where our 'wood field' came to an end. On a glorious August day we were re-making the 'stooks' of corn in this field. God's yellow furnace was blazing fiercely in a clear blue sky and a gentle breeze was rustling the ears on the sheaves. The day was perfect but the work was hard. Despite the heat I had changed my sandals for wellington boots because the sharp stubble had cut and scratched my ankles. This was a field of oats for the horses that Mick had cut in the traditional way using a scythe. Mick was good with the scythe. With each swipe of the long curved blade he would throw out a sheaf-full of corn, which Daddy would bind together using a fistful of strong stalks. The sheaves were put leaning against each other in stooks in straight rows across the field and a few weeks later these had to be re-made into bigger stooks.

Mammy had brought the four-o'clock tea and had stayed to help. We, as children, were expected to do our part which meant taking the sheaves from two stooks and building them into one larger stook that was then 'headed' with four sheaves on top, turned upside down. This allowed the grain to dry thoroughly for threshing later on.

I'm helping Mick. My seven-year old arms are aching from lifting the heavy sheaves. My face is stinging because I have wiped the sweat off with my stained hands and the sap of the red-weed is burning my skin. Suddenly, Mick has stopped working and is gazing off into the distance. My eyes follow his gaze. Mary Flaherty is coming across the field in our direction. She is wearing a summer dress with a bright flowery pattern and a big bonnet to protect her from the sun. Mick's greeting is jovial and he makes some flippant remark about her bonnet. She isn't suitably dressed for the work we are doing but nevertheless joins in with myself and Mick. That's when the fun begins. Even though they are both well into their middle years they spend the rest of the day messing like teenagers. Sheaves of corn are being thrown about and Mick 'accidently' knocks Mary to the ground bringing the whole stook down on top of her. I forget about my stinging face and aching arms and join in the fun.

Circumstances and societal norms had condemned them both to lives of chastity. He was a bachelor without prospects and she was a spinster without a dowry, but I knew nothing of such things. I kept encouraging him to knock her down again. It was a great distraction and broke the monotony of the work. Mary had brought some excitement and laughter into our day with her funny stories and witty remarks. Even though the work rate had slowed down, Daddy made no comment. A strong bond existed between himself and Mick; a bond that could never be expressed in words, only in deeds.

Mick was the older brother and this seemed to colour their relationship. Daddy would always let Mick 'win' at whatever work they were doing together. Mick's ego demanded that he should finish the job ahead of everyone else, whether that was thinning turnips, or piking hay, or whatever. One chilly Saturday in early spring, when I was six and Kevin was seven, we were removing spuds from their long pit in the 'front lawn' field. The potatoes were stored through the winter in long pits, in the garden where they had grown. These pits were covered with a thick layer of straw, on top of which was an even thicker layer of topsoil to protect them from winter frost. In springtime, when nature called, the potatoes would begin to sprout shoots and had to be taken into a shed for the remaining few months until the new potatoes were ready. I was helping Daddy and Kevin was helping Mick. The work meant breaking off the sprouts as we filled the potatoes into the butter-boxes. These butter-boxes were large wooden boxes that were very useful pieces of equipment in all farms during the fifties and sixties. When the butter-box was full it would be emptied into the horse's car and when the load was full would be taken into a shed. Watching Mick, I must have sensed the competition because I was working all out and Daddy and I were winning. We were filling and emptying faster than they were and then for some unknown reason Daddy slowed down and they were winning. I remember crying as I urged him on, but to no avail. Kevin was jeering and telling us to hurry on, but Daddy had slowed his pace and I could not get him going again.

For reasons beyond my comprehension, Mick resented the presence of Lizzie and Sara in the house. He seldom spoke to them and never in friendship. Sometimes he would stare at them from across the kitchen with fiery rage in his eyes. The ill-feeling was palpable. It was deep-seated and constant but never vocalised. Later in his life he would pay a high personal price for this anger but for now Mammy was there as a neutral member to diffuse the unpleasantness. Mick held her in very high regard and loved her in a special way. There was nothing he wouldn't do for her and outwardly showed more affection and caring than her husband did, but his relationship with his two sisters was a different matter. The unresolved bitterness was a blight on his life. Shortly after Lizzie's sudden death in 1960 Mick went into decline and suffered a mental breakdown. He lived on for ten years but he was lost to us; that wild funny man that we admired and loved would sit in the corner fidgeting with his jumper and muttering to himself, unable to focus on anything; a stranger amongst us.

Tom Leen came to visit one day and tried to engage with him. He had brought an old sugán stool that needed repair. He sat for hours with Mick encouraging him to squeeze the twine tightly back and forth across the rungs. Like an obedient child Mick complied but his limp hands were just going through the motions. His frightened eyes never left Tom Leen's face but no words came from his mouth. It was no use, Mick was beyond helping. He was to endure umpteen sessions of electric shock treatment up to his death. Mammy took him to Tralee every few months for this barbaric treatment and on one occasion to Dublin where some new treatment was being tried out. It made no difference; it only added to his private hell.

From a very early age I was told that I bore a strong resemblance to Daddy. It thrilled me to hear neighbours say to him "He's the image of you". I think this caused me to develop a special bond with him. After school I would walk to wherever he was working to see what he was doing. I remember taking the four o'clock 'tay' to him one evening when he was ploughing at the far end of the 'back lawn' field. I might have been five or six at the time. The tay consisted of a bottle of hot sweet tea and four thick slices of home-made brown bread. Mammy had put the bottle into an old woollen sock which retained the heat and made it easy for me to carry. He saw me coming from a distance and pulled the horses, Buck and Moll, to a stop at the far headland and sat down by the ditch.

I trudge along slowly. The whole headland is yellow with dandelions, 'showing their unloved hearts to the world'. My heart feels a little sad too because Daddy seems so all alone and lonely in this secluded part of the big field. I make up my mind to stay. As he sits there drinking the tea, he points out the different birdsongs we're hearing, saying "That's a thrush, that's a blackbird, that's a robin". I can only distinguish the cawing of the crows as they scratch for worms in the upturned earth. He lists the names of some trees and bushes growing along the ditch and tells me what uses can be made of the different kinds of wood. Blackthorn saplings could be made into fancy walking sticks, but ash is the best for hurleys or tool handles and sally scallops are used in the thatching of houses. My head is addled, it can't take in all the information it's receiving.

Daddy had acquired a deep appreciation of nature and elementary science from long years in Nature's school, but his formal education only ever manifested itself as one poem; the poignant tale of Caoch the Piper. The pathos in this poem appealed to him and he would often quote a few lines to suit his mood. It was a poem set in former times and seemed to have some hidden meaning for him.

> *One winter's day long, long ago, when I was a little fellow,*
> *A piper wandered to our door, grey headed, blind and yellow.*
> *And oh, how glad was my young heart, though earth and sky looked dreary,*
> *To see the stranger and his dog, Pinch and Caoch O'Leary.*
> *And when he stowed away his bag, cross-barred with green and yellow,*
> *I thought and said, "In Ireland's ground there's not so fine a fellow."*
> *And Fineen Burke, and Shaun Magee, and Eily, Kate and Mary,*
> *Rushed in with panting haste to see and welcome Caoch O'Leary.*
> *O God be with those happy times, O God be with my childhood,*
> *When I bareheaded roamed all day, bird- nesting in the wildwood.*
> *I'll not forget those sunny hours, however years may vary.*
> *I'll not forget my early friends, nor honest Caoch O'Leary.*
> *Poor Caoch and Pinch slept well that night, and in the morning early*
> *He called me up to hear him play, "The Wind that Shakes the Barley:"*
> *And then he stroked my flaxen hair and cried, "God mark my deary"*
> *And how I wept when he said "Farewell, and think of Caoch O'Leary."*
> *The seasons came and went, but Old Caoch was not forgotten,*
> *Although we thought him dead and gone and in the cold grave rotten:*
> *And often when I walked and talked with Eily, Kate or Mary,*

We thought of childhood's rosy hours and prayed for Caoch O'Leary.
Well twenty summers had gone past and June's red sun was sinking,
When I, a man, sat by my door, of twenty sad things thinking.
A little dog came up the way, his gait was slow and weary,
And at his tail a lame man limped -'Twas Pinch and Caoch O'Leary.
Old Caoch, but oh how woebegone! his form is bowed and bending,
His fleshless hands are stiff and wan, ay, time is even blending
The colours on his threadbare bag; and Pinch is twice as hairy
And thin and spare, since first I saw him, trailing Caoch O'Leary.
"God's blessing here!" the wanderer cried, "far, far be hell's black viper:
Does anybody hereabouts remember Caoch the Piper?"
With swelling heart I grasped his hand, the old man murmured "Deary,
Are you the silky-headed child that loved poor Caoch O'Leary?"
"Yes, yes," I said—the wanderer wept as if his heart was breaking—
"And where, avic-machree," he sobbed, "is all the merry-making
I found here twenty years ago". "My tale," I sighed, "might weary:
Enough to say there's none but me to welcome Caoch O'Leary."
"Vo, vo, vo!" the old man cried and wrung his hands in sorrow:
Pray let me in, astore machree, and I'll go home tomorrow. My peace is made,
    I'll calmly leave this world so cold and dreary;
And you shall keep my pipes and dog and pray for Caoch O'Leary."
With Pinch I watched his bed that night; next day his wish was granted,
He died and Father James was brought and the Requiem Mass was chanted.
The neighbours came to dig his grave near Eily, Kate and Mary.
And there he sleeps his final sleep--God rest you Caoch O'Leary.

But now is not the time for poetry, there's more ploughing to be done. When he's finished the tea and bread he makes a cushion of his coat so that I can have a ride on the plough. I sit there, perched precariously behind the two wheels watching the slow measured movement of the horses' legs. Instinctively, I know that Daddy is good at this work. With minimum words of communication, the horses know exactly what to do and they maintain an even, steady pace from one end of the field to the other. Moll is on the grass sod while Buck's big hooves have to fit into the narrow furrow between the upturned soil and the grassy verge. I'm fascinated - watching the blade of the plough cut through the green sod and turn it over so neatly. I can feel the effort Daddy is making to keep the plough steady on the rough ground. For me it's all very exciting even though it's a bumpy ride. At no stage does he say 'that's enough for now' but patiently continues putting up with the inconvenience. The field is stony and my bum is feeling sore but I stay because he tells me I'm doing a good job keeping the plough steady. The ploughing continues till dusk sets in. He unhitches the horses; lifts me up on Buck's broad back and holding the 'winkers' of the two submissive animals, plods his weary way homeward. When the horses are given the freedom of their own field, they lie down on the grass and roll over and over on their backs with their legs flailing the air. It looks like an expression of sheer delight to be free of their harness. I stand with Daddy at the gate and watch them go to the stream where they gulp in a large quantity of water and then start grazing. Tomorrow they will have more work to do.

On a cloudy day in late spring a Ford Prefect car drove into our yard. We gazed in amazement as a tall man in black stepped out wearing a hat and a scarf. His rimless spectacles were perched high on the bridge of his pointed nose, giving him an air of superiority. His shoes were glossy black, matching the leather gloves on his hands. He was elegance personified. In a loud accent that was strange to our ears, he called out "Hello Madge" as Mammy rushed to greet him. He was her uncle, Fr. Dan O'Hanlon, home from America on a holiday, paying a surprise visit. She escorted him into the parlour and quickly put an action plan in place. Kevin was sent to the shop to buy a cake and biscuits and I was sent to find Daddy. I found him in the piggery yard repairing a stone wall that had been broken down. His clothes and his shoes were covered with dirt. I told him Mammy wanted him to come into the parlour to meet the priest from America. His first reaction was negative but after a little while he changed his mind. He washed his hands in the tar barrel by the hayshed and told me to go up stairs into his room and to drop his 'Sunday suit', his new cap and his shoes out the window. He changed in the hayshed and went in to greet the visitor, looking like a gentleman farmer. With Daddy as anchorman, trying to talk posh in the parlour, Mammy went off to make the tea, using the fancy teapot and the delicate china tea set she kept for such occasions.

So many fond childhood memories of this tall, solitary man still pluck at my heart strings but two disquieting memories keep returning. On an autumn evening, when I was nine, a stray sheep-dog, with weary gait and sad eyes, ambled into our yard. His thick black and white coat was matted and dirty. The poor auld thing was in bad shape; he obviously had been roughing it for some time. He looked furtively from left to right, unsure of his welcome. I called to him but he didn't move, so I walked slowly towards him making friendly sounds. He retreated a few paces and then stopped; his faith in humans was low. I ran to the kitchen and got a slice of buttered bread from the press and with this lured him to a safe place at the back of the turf-house. I made a makeshift little house for him in a sheltered corner of what was once the main hall of the Great House. Tonight he could sleep with the ghosts of the upper classes, but deep down I knew he would not be welcome in the farmyard. So for many days I kept him hidden, tied with a piece of rope, in this secluded corner and secretly supplied him with food and water. Whenever I had time I sat beside him and brushed his dirty coat with an old scrubbing brush. He spent most of his time sleeping and recuperating and I could see he was getting better. Then one day, when he had his strength back, he chewed through the rope and ventured into the yard.

Daddy always kept a 'good' dog for the cows so when he saw this stray he hunted him out the gate, chasing him down the road with stones. But the dog had been shown a kindness so he re-appeared the following day only to be driven off again. Alone and unloved he kept returning till one day Daddy tied him to the wheel of an old farm machine and beat him severely with a heavy stick. Blood was oozing from his nose and mouth and there was a livid gash over his right eye. I watched in horror from a distance as he yelped in pain. But it didn't end there; he was beaten repeatedly before being let loose. The tortured animal ran yelping out of the yard and down the road, his pleading whine echoing in the recesses of my brain. A confusion of feelings welled up inside me. The father I loved and worshipped had done a terrible thing to a poor dumb animal.

But the dog came back! Some days later Daddy found him hiding in the little bed I had made for him behind the turf house. He went into the kitchen and got the keys of the Morris Minor from the drawer in the dresser. He called Kevin and told him to get the dog into the backseat of the car. When I saw what was happening I jumped into the back seat too. Daddy started up the car and drove a long distance in the direction of Clandouglas. I didn't ask any questions but Kevin, who had also grown fond of the stray, knew what was happening. He kept feeding the dog from the packet of 'marietta' biscuits he had in his pocket and patting his head. Along a lonely stretch of the road Daddy stopped the car. He opened the back door, pulled the dog out onto the road and then drove away. I watched my poor old dog through the rear window; a forlorn and bewildered little figure that began to grow smaller and smaller. The car went around a bend in the road and he disappeared from sight.

Some weeks later we were drawing home the turf from the bog at Ballinagar. Daddy had hired a local contractor with a tractor and trailer and had employed three workmen for the day. Each trailer load of turf would be tipped-out in front of the turf-house to be thrown in sod by sod through the large arched opening of what had been the rear entrance of the Great House. It was a gruelling day's work so on the last trip from the bog the driver pulled to a halt at JJ's pub in Lixnaw village. Daddy was in a hurry to get home and was totally opposed to the idea but didn't get his way, so he remained outside as the men trooped in to the bar. I hesitated for a moment but Uncle Mick was going in so I went with him, already tasting the refreshing tingle of the red lemonade in my throat. While the men were drinking their pints of Guinness one of them made up a mixture of stale, watery Guinness and picked on me to be the Judas. I must have known there was malevolent intent in the workman's practical joke. Why I went along with it I do not know but I brought the mock drink to Daddy as he waited by the tractor. He took the glass from my hand, emptied the contents on the ground and gave it back to me without saying a word. Our eyes met for one brief moment and I knew immediately the wrong I had done. I slinked back into the bar feeling guilty, ashamed and confused. I sat quietly in the corner with my lemonade as the men guffawed at the counter. I knew I had been used and I hated them for it. I was angry with myself and I was angry with Mick for letting it happen. The journey home was silent. The load was tipped-out in front of the turf-house and the men received their wages, but no words or handshakes were exchanged. Daddy would never ask them to work again. My day had been ruined. The bad feeling would not go away. I wanted to say sorry but didn't know how. Instead, it got locked in my memory. I felt I had betrayed the person I most admired in the whole world.

Chapter Five

Our Animals

Two strong horses were central to most of the work that went on in our fields. They were wintered in the well-designed stables that had seen many horses come and go over three hundred years. Overhead the stables was a large hayloft. Every summer this loft would be filled up with hay that had to be piked in through the front window. During the winter the hay would be thrown down through an opening by the wall into the long manger where the horses were haltered. As children we enjoyed doing this and would be rewarded with a gentle neighing sound from the three animals down below. The three animals were Buck, a huge jet black gelded shire that occupied the first bay, Moll, an Irish draught chestnut-coloured mare, in the middle bay and the white pony filled the third bay. These were Daddy's pride and joy, specially selected with his keen eye for a good horse. But one autumn Buck got sick and couldn't be cured. The vet believed that he had eaten the ragworth, a poisonous weed that caused a slow deterioration of his brain and his powerful body. He lived out his few remaining months within the walls of the big orchard. Most of the apple trees in this orchard had died off but some remained and were dropping their ripe fruit on the ground. As well as fruit there was an ample supply of lush grass but poor old Buck had no appetite. Most of the time he stood by the wall with his head hanging low. At other times he would behave in a very strange way. He would attempt to raise himself up by propping his front legs high against the wall as the goats did when trying to reach the ivy. Perhaps he felt the ivy held some magic potion for his throbbing brain. It was both mystifying and frightening to observe this weird behaviour by this huge skin-and-bone animal. And then one morning he was dead. Daddy and Mick dug a deep hole at the corner of the orchard and Moll, his stable companion, was used to drag the huge black body into it. Moll was very skittish and difficult to control and Daddy had to take her away immediately, leaving Mick to fill in the grave.

Chapter Five

Some years later, another tragedy cost the life of the white pony. The pony's role was to pull the trap to Mass on Sundays or to take Mammy to Tralee or Listowel for shopping. But on one particular frosty morning, with black ice covering the road, Mick decided she would be best for the creamery run because her shoes had been fitted with ice studs. She looked a bit small between the long shafts of the creamery car but she was a spirited animal and had a stout heart and set off at a steady pace. All went well until the last stretch of the homeward journey. When she was coming down Neenan's hill she skidded and fell on her knees and got pushed along in this position over the rough icy road. Mick was powerless to help. When the momentum ceased he gingerly got her to stand up and surveyed the damage. It was bad; the skin on her legs was shredded to pieces with bloodied bone exposed and both kneecaps bleeding profusely. He got her out from between the shafts and gently coaxed her to walk home. The vet was called but he said there was nothing he could do; there was too much damage to the knee joints. She stood shivering in the stable for twenty four hours unable to move, her wounded legs oozing a dark sticky fluid. Then a lorry came and took her away. Our gentle little pony was gone. She was replaced by a new horse, a 'cob' called Paddy. He was smaller than a plough-horse but larger than a pony and was suitable for pulling the trap as well as doing work in the fields. He was a good horse but highly spirited and I was always a little bit afraid of him.

Of all the horses that were to come and go through our boyhood years none would make a bigger contribution to the family welfare than Moll. Daddy loved this animal more than life itself. She was a perfectly formed Irish Draught mare with a dark brown coat and a willing nature. For some reason or other he never had her shod even though she made regular long trips to Tralee or Listowel with a load of pigs. Every second year she produced a foal that provided additional income when sold as a two-year old. All her foals were beautiful but one in particular was a rare gem. He was a male foal with a blond coat and a sturdy build. He was delightful to observe and as he grew older everyone remarked on his beauty and perfection as he raced about the field. He was still suckling and the bond between mother and foal was strong. Moll had been taken away from him to pull the hay-float, to get the hay in from a nearby meadow. The mare and foal called to each other all day long and as evening approached we could see the milk spouting out of her dark teats. Daddy milked her onto the ground to relieve some of the pressure on her swollen udder. From the farmyard she whinnied loudly and the foal responded from his enclosure back in the field. The urge to be with her was so strong that he leapt over the fence into a part of the field that was cordoned off with barbed wire.

Unseeing and unaware he ran at full gallop into the double strand of the sharp barbed wire, shredding his skin to pieces along the left side of his body. Limp bits of flesh were hanging loosely as the poor stricken creature made his way to greet his mother. As he suckled her udder blood was dripping from his many wounds. A vet came and tried to patch him back together but the cuts were deep and numerous and the rusty barbed wire caused septicaemia. He was put in a secure place in the front stall but he didn't survive very long. He lay shivering in his prison cell unable to rise; his beautiful fawn coat matted and dirty. Moll could be heard calling to him from the field and he would give a low feeble response, but each day he grew weaker. He was buried near Buck at the corner of the orchard. Moll gave birth to many other foals over her lifetime but none were quite so beautiful.

The affinity between Daddy and this animal intensified over the years and on those long lonely trips to fairs, horse and man communicated in their own special way. "Go on, Moll" was the only encouragement she needed to pick up the pace. Hitting her in any way would have been anathema to him. As the years rolled by times began to change; traditional style farming was giving way to modern machinery. Daddy was growing less able to do the hard work so he made over the farm to Kevin, and had to part company with his beloved Moll. She was replaced by a Massey Ferguson tractor. Daddy stood at a distance as a lorry came to take her away; this animal that had served him so well for so many years was destined for the knackers' yard. His sad face and wan gaze followed the lorry down the yard as Moll, with a low whinny, disappeared out of his view and out of his life. Without a word he turned and headed back through the fields, limping as he went; his right hip-joint was getting worse and causing him pain. We saw his tall lean figure disappear behind the orchard wall. It was late in the evening when he returned and his mood remained subdued for many days to follow.

***

During spring, summer and autumn, the hand-milking of the cows was an ever-present chore around which everything else had to be planned. Trips away to town, to hurling matches, or to the seaside would always be brought to an end with the phrase:"Time to go; we have to be home for the cows". They were our masters, we were the servants. We slaved all year to ensure their wellbeing through the winter and they lounged around the fields all day stuffing their bellies with lush green grass and then lying down contentedly to chew the cud. Work in the meadows and gardens always came to an end with a familiar call: "Lads, will one of ye go for the cows". I liked the sound of that call because it meant getting early release from the work we were doing. The cows were always glad to be rounded up and I used to feel a bit sorry for them as they lumbered along with their swollen udders making it difficult for them to walk. In spite of this they would sometimes try to run, tails up, when the warble fly was about. Some in-built radar warned them about this nasty insect. It would lay its eggs in the cow's leg. When hatched out, each larvae would tunnel its way through the muscle and flesh until it reached the cow's back. There it would morph into the pupa stage and you could feel them as a series of lumps along the backbone of any cow that had been infected. Daddy would pour on some kind of insecticide to relieve their agony. While these parasites didn't affect the milk they greatly reduced the yield of infected cows and reduced the value of the hide when the animals were slaughtered. Many hands were needed for the milking and so from a very early age we were all taught how to milk a cow and were expected to fulfil our milking duties both morning and evening without fail. This usually meant milking one or two cows each because the grown-ups would have two done while our small hands struggled with one. All our cows had names and each had their own individual personalities. They were all of the Shorthorn breed, which was the most common breed of cattle in Ireland before the introduction of the Friesians. The Shorthorn was a versatile, hardy animal that came in a variety of colours.

Curly was a Shorthorn cow of rare beauty. Her soft curly hair was a mixture of white and deep burgundy and her horns were two perfectly-shaped arcs above her head. I had learned to milk on Curly because she was quiet and released her milk freely. Sitting on a little three-legged stool with my head against her flank, I felt a strong bond with this huge gentle animal. For some of the other cows, one required very strong finger muscles to draw the milk, so I had staked my claim on Curly. In the summer months of lush grass she would fill the bucket to the top. I loved hearing the 'pluuish-pluuish' sound of the jets of milk hitting the frothy top as the bucket filled up. Sometimes, to relieve the boredom, I'd aim a jet of milk at one of the cats who always hung around at milking time, and, while they hated being shot at, they loved the taste of fresh milk. Other times I'd aim a jet of milk at Tom or Kevin on the opposite side of the stall and then play the innocent, in the full knowledge that the response would be immediate and deadly; a jet of warm milk would be aimed at my head. But wasting milk would draw the wrath of both parents if we were caught, for this was the main source of income for the family. The creamery cheque, issued at the end of each month, was the justification for our enslavement to these animals.

Brindle and Big Spot were the leaders of the herd. They both competed for the number one slot in the pecking order. Both were big animals and were well formed with a perfect set of horns. Most of the cows had crooked ill-shaped horns but these two had long pointed horns that curved upwards just like *el toro* in the *Corrida de Toros*. They were beautiful, majestic animals. One evening as they approached the stall door Brindle lunged at Big Spot goring her in the flank. The vet was called and he stitched up her wound but that was not the end of it. After the milking was done both were kept in the stall with their heads secured in the wooden frame. Like two disruptive school children they awaited their fate -- dehorning. The vet approached them with what looked like a huge hedge clipper with very long handles and sharp blades. The cows struggled backwards but there was no escape. The blades sliced through the base of the horns exposing a red raw hole as the horn fell to the floor. Blood began to spout but he poured some liquid on it to make it stop. In Brindle's case he had cut too deeply into her skull at the left side and had opened an artery. The blood didn't just flow out, it squirted out as if from a water pistol. In her pain she kept shaking her head, spray painting the stony wall with artistic blood spatters, where I had split my forehead as a four year old. I watched her stand there with her head hanging low, just like old Buck had done in the orchard before he died. The vet was concerned about the loss of blood so he wrapped a long white bandage around her head many times and said to leave it on for a day or two. The blood oozed slowly to the surface, creating two red patches where the horns had been. Brindle survived the ordeal but had lost her noble look. With her bandaged head she looked like one of the Boer war veterans in the Kitchener book that Mammy kept in the kitchen drawer.

Every year the whole herd would have to be tested for tuberculosis. The vet would jab a needle into the neck of each animal and come back a week later to analyse the results which were interpreted in three ways; positive, negative, or inconclusive. Those that proved positive had to be sold, 'inconclusives' would be given a reprieve for one more year. Bess, a fawn-coloured cow, never failed the test. She was Mammy's favourite and she had always been TB free, so her milk was kept for household use. All through spring, summer and autumn a tin gallon was filled to the brim with Bess's raw milk for our use throughout the day. Whatever milk was left over Mammy stored in a large jug, where it turned sour and was used for baking. If Daddy was feeling very thirsty he would drink the sour milk. He couldn't drink Guinness unless it was heated and sweetened with sugar but he could drink the sour milk even when it had become curds and whey. But during the winter months we had to use the milk sparingly. All the cows were dried off to allow for gestation but one cow would not be put in calf and would not be dried off. She was called a 'stripper' and would keep producing a small quantity of milk all through the winter.

Every morning and every evening the long line of cows would be led in by the two goats and bringing up the rear would be the bull. We believed that the bull posed no threat as long as he was with the cows. But in the autumn when his services were no longer needed he was put into solitary confinement in the big orchard. Just inside the orchard gate there were two old pear trees that produced huge crops of soft juicy sugar-pears. If the bull was grazing at a distance, we would risk going in and knocking down the pears with a long stick. This would usually attract his attention and we would have to make a hasty retreat. One evening John Joe Whelan was looking at the yellow pears by the orchard gate while the bull was on the other side staring at us. John Joe loved those soft sweet pears even more than we did. So 'macho man' climbed over the gate and with a loud shout ran at the bull. The bull took fright and ran away but as John Joe retreated he heard a loud bellowing behind him as the bull prepared to charge. Like an Olympic sprinter, John Joe dashed to the gate and barely got over, as the bull crashed his horns against it. It was a close call. John Joe was visibly shaken by his narrow escape and stood there cursing the bull. The enraged animal continued pawing the ground with his right front leg and bellowing loudly. There would be no juicy fruit this day. As we turned to go, the leaves of the pear tree rustled and the sweet yellow fruit glistened teasingly in the evening sun.

And then the A.I. service was set up in Castleisland and we didn't need the bull any more. Science had replaced him. Artificial insemination was seen as the way forward for improving livestock. Whenever a cow was 'bulling' the 'man in white' would be summoned. When all the other cows were let out in the morning the cow in question would be kept behind in the stall, awaiting this undignified experience. The 'man in white' would put on a long rubber glove that reached above his elbow and reaching deep within her body would deposit a quantity of semen. When he was finished the job Mick would bring him a bucket of water for washing and they would make some unsavoury jokes which we were not supposed to hear. Around that time farm advisers were encouraging dairy farmers to breed Friesian cattle because they gave a better milk yield and because the bull calves were quicker to fatten. So gradually our herd of cows began to change colour; from multi-coloured Shorthorns to black and white Friesians.

The two goats produced a small quantity of milk which was kept for use in the kitchen also. These two goats produced two kids every second year. There was no AI for them; the old-fashioned way was good enough. Daddy kept the goats with the cows in the belief that the presence of goats protected the herd from bovine tuberculosis and other diseases. Cows are always milked from the side but goats are milked from the rear. Our two goats were very tame animals and would stand quietly to be milked anywhere in the yard. This was very amusing to watch and if we were laughing while milking them, they would turn their heads to look at us, as if to say get on with the job. They usually gave birth in early spring in some secluded corner of the field.

One night Daddy was home late from his visit to the 'rambling house'. This was Flaherty's house at Crotta Cross, where Mary and Frank Flaherty, brother and sister, kept open house for all and sundry to drop in and chat on long winter evenings. It was pitch dark when he got home this night and myself and Kevin were playing ball at the end of the kitchen. He gave his flash-lamp to Kevin and suggested we go down to the end of the orchard wall to see the two new-born kids that Beth, the older goat, had given birth to. My fear of the dark was far greater than my desire to see the new arrivals but Kevin had moved towards the door and so I dutifully followed. The flash-lamp only gave a small circle of light on the rough ground but after a little while our eyes began to adjust to the darkness.

As we walk down through the yard the ruins of the old Great House stand out stark and menacing against the moving clouds. The wind whistling through the giant conifer trees sounds like the re-awakened ghosts of long-dead Ponsonbys. A shiver goes down my spine. I want to go back but Kevin is holding the flash-lamp so together we move on towards the huge door that opens into the orchard. My fear of the gander returns but is pushed aside by the paranormal shapes of the trees that rear up in front of us, their bare skeletal branches reaching out to grab me. What would have been such a pleasant scene that morning is now such a nightmare. We pick our way gingerly under the apple trees, getting poked in the head by low-hanging branches. We search here and we search there, calling to Beth but to no avail. Just as we are about to give up we hear a short 'meighhhh' and there she is in the corner, behind a clump of briars, with two delightful little creatures suckling away to their heart's content. Being in the presence of the goat with her kids brings some normality to the situation and temporarily allays my fears. Beth seems at ease and not the least bit frightened of the ghosts and ghouls that wander about these ancient walls late at night. Then we decide to go back. But which way is back? We are both very disoriented from our search. Serious panic begins to set in as the flash-lamp dies. We are Hansel and Gretel lost in the woods of apple trees. Kevin moves in what he believes is the right direction but stumbles and falls and I, following behind, fall over him into a puddle of foul-smelling stagnant water. I'm on the verge of tears when suddenly a small white shape appears out of the darkness. It's Rose, our faithful sheepdog, who has come to our rescue. She has followed our scent down to the orchard. Kevin grasps her collar as she leads the way back and we are soon within the warm glow of the kitchen fire. Our wet and muddied state causes initial shock which turns to laughter as we relay our story.

As they got older and filled out, Beth's two kids were a delight to look at and play with. All they ever wanted was to frolic around and have fun and they loved the attention we gave them. I thought this must be the reason why children were referred to as 'kids'. We loved to chase after them and when we caught them they would nudge their little heads against us, sometimes a bit too hard, especially when their tiny horns began to grow. Spring was in the air and they were thriving on their mother's milk but their wonderful happy existence would be cut short. They were surplus to requirement and we witnessed their horrible execution. Daddy suspended them both by their hind legs from a beam in the shed and with his pocket knife made a small cut on their stretched necks and allowed them to bleed to death. Their pitiful cries were heartrending and the free-flowing blood was horrifying to watch. Then it stopped flowing and the two lifeless carcasses were skinned. The two delightful creatures were gone and all that was left was meat and two wet slimy hides. We nailed the hides, stretched wide, on a south-facing door so that they would change into leather. Goatskin leather was used for making *bodhráns* (drum), so a few months later I tried my hand at *bodhrán* making, but with little success.

Our two goats, Beth and Meg, were a mother and daughter pairing. They were sometimes neck-tied together with a piece of rope about two feet long that restricted their movements and prevented them from climbing. Goats are nimble-footed climbers and their favourite food is the ivy leaf. Over the centuries the orchard walls had been colonised by a heavy crop of lush green ivy. It covered the tops of the walls all the way round the five acre enclosure, giving the walls a beautiful artistic finish. Sometimes when fodder was scarce Mick would climb up on the walls, slash hook in hand, and cut down the ivy leaves for the 'dry' cattle that were wintered there. One winter the goats were confined to the small orchard and had picked the ivy off the side of the walls as high up as they could reach standing on their hind legs. The temptation to reach the ivy higher up the wall was too great, so Beth and Meg attempted to climb a slanted plum tree growing by the wall. But things went horribly wrong. One must have slipped causing the other to lose balance. They fell on either side of a strong branch and, choked by the rope, they died together, suspended in mid air.

***

References to pigs in the kitchen were often used to create a 'stage Irish' image of a rural Ireland in less sophisticated times, but sharing the kitchen with a farrowing sow was seen as perfectly normal to us and was a necessary practice in animal husbandry. Luckily our kitchen, thanks to some bygone Ponsonby planner, was huge. When a sow was ready to give birth Daddy would cordon off an area by the back window with a low wooden partition. A bed of straw would be put in place and there the sow would give birth to her litter of pinky-white *banbhs* (piglets). The birthing of a big litter of *banbhs*, usually about a dozen, was a slow process and it was Mammy's role to act as midwife, sometimes all through the night. The life of each little *banbh* was precious in terms of pounds, shillings and pence and so on one occasion, when one little fellow was so small and weak that he could not survive the rough and tumble of the litter, she put him to one side in a cardboard box to be reared as a pet. She called him Pigeen and we took turns feeding him using a baby's bottle and teat. It was so exciting and rewarding to see his eager little mouth grip the teat and gulp down the milk.

The rest of the litter suckled greedily on the sow and would pinch her teats as their teeth began to grow. This would cause her to jump up suddenly with potential danger to some of the litter, so Mammy would break off their teeth with small pincers; cruel but necessary. The sow and her other eleven offspring remained in the kitchen for a week or so until the *banbhs* were hardy enough to survive in the piggery. Pigeen's growth was retarded and so he remained in a corner of the kitchen in his cardboard box. Like any pet animal Pigeen grew to love his human surroundings and roamed freely about the kitchen as a dog or cat might do, nudging his snout against our feet at the dinner-table. When he was seven or eight weeks old his privileged status came to an end and he was put into the piggery with the rest of the litter who were now much bigger than him. He fared alright in this new environment but never lost the bond with us, his carers and playmates. As they grew older they were allowed free range in one of the fields. The sow, a large White York, loved to range across the field eating grass and painfully attempting to root up sods of earth. To prevent sows from ploughing up fields with their snouts, a cruel but effective measure was used; a number of steel rings were pierced through the tip of the snout.

Daddy was expert at 'ringing' sows, using a special type of pliers designed for the job. So before White York could be let out in the field she had to endure the ordeal of 'ringing'. I was his helper on this occasion. He made a noose at one end of a piece of rope and backed the sow into a corner of the piggery, where she stood cowering and whining. He slipped the noose into her open mouth up past the two large tusks on her upper jaw. Squealing and resisting, she was dragged to the strong iron gate and the rope was tied to one of the bars. The frightened sow unwittingly secured herself in this position by pulling backwards on the rope with all her strength. Daddy then took off his peaked cap and told me to hold it over her eyes so she could not see what was happening. Deftly he inserted four steel rings along the rim of her snout. White York squealed loudly as he put each ring in place, but his steady, accurate hand caused no blood to flow. White York would have the freedom of the field but without the pleasure of rooting up sods and searching for grubs, bulbs and other tasty morsels. Her litter of *banbhs* followed her around the field with little Pigeen bringing up the rear. They stayed with her for another month or so and then were weaned off and fattened on pig ration and were ready for sale in ten or twelve weeks.

One gruesome experience, however, marked the 'rite of passage' to the weaned state for male *banbhs* - castration. As luck might have it, Pigeen was male and, even though much smaller, had to endure this painful operation. Daddy placed an old chair in the corner of the piggery yard on which Mick sat holding the *banbhs* upside down on his lap, with a firm grip of their four legs. Myself and Kevin chased the unwilling patients, grabbing each one by the tail and bringing him to the operating corner where Daddy, with a surgeon's skill, made two small incisions on the scrotum. The removed testicles were dropped into a jam-jar and later thrown to the cats. Pigeen was the last to be caught. He was hiding inside the low roofed building where they slept at night. Our piggeries were a legacy from the past, and were very well laid out with an indoor and outdoor area for each litter. It was generally believed that Kitchener (Senior), who was reputed to have taken a hands-on interest in upgrading his newly acquired farm, had these piggeries built to the latest design. It was dark inside but I managed to catch him by the hind leg, saying sorry to him as I lifted him up in my arms and brought him to Mick. It was over in a matter of minutes. Pigeen didn't squeal much but he would sing in a high castrato voice for the rest of his short life.

The rearing and fattening of pigs was an important source of additional income, especially over the winter months when the monthly cheques from the creamery had ceased. *Banbhs* and pigs were sold at weekly markets in Tralee. Daddy would sometimes take one of us with him on such trips. My turn came when Pigeen's batch were ready for sale. He woke me at 5 a.m. and we had a quick breakfast of tea and brown bread. I helped him put the harness on the horse; winkers, collar and hames, straddle and britchen, the full rig-out for a horse to carry a heavy load, up and down hills, the nine mile journey to Tralee. Moll, his beautiful chestnut mare, was always chosen for this journey because she could make faster time. He guided Moll and the high-railed cart to the piggery gate and I, being the more agile, went in to catch the *banbhs* one by one and bring them to him for loading. The trick I learned was to catch the tail with my left hand and lift quickly by putting my right hand under the chest. This rendered the animal immobile and eased the loading process. Pigeen, who was still noticeably smaller than the others, was the last to be loaded.

The first pale streaks of dawn are brightening the eastern sky as we begin our journey. Both of us are sitting high up on a plank of wood that Daddy has placed across the top of the rail. As we pass our house, I gaze at the dark windows to see if anybody else is up and wish I could be back in my bed, snuggling under the blankets. A few guttural words are directed at Moll and she responds by picking up the pace, while beneath us there is a continuous chorus of grunts and squeals from our discontented passengers. Daddy asks if I'm feeling tired from the lifting and then poses a semi-philosophical question: If I lifted the same banbh every morning from when it was little, would I be able to lift it when it was a fully grown pig? I ponder on that for a while as I sit there looking down at Moll striding along Garrynagore road with her long mane blowing in the wind. Silence descends upon us. The banbhs have settled down and the rhythmic clip-clop sound of Moll's hooves induces a hypnotic state in the stillness of the early morning. My eyes are beginning to close when the silence is rudely interrupted by the loud barking of Lawlor's dog, who proves his worth as a guard-dog by following along behind us until we are well away from his house. As we turn the corner at Abbeydorney village an enormous orb of golden light is slowly rising above Stacks Mountain and painting the countryside in dappled colours. It shines directly in our eyes, making it difficult to see the road but Moll needs no guidance. The hardest part of the journey lies ahead as we begin the tough climb of the Short Hills. Moll

Chapter Five

is sweating now and breathing hard but a few muttered words from her master drives her forward. My admiration for this magnificent animal is growing; with her strong shoulders pushed hard against the collar, she never slackens her pace until we arrive at Tralee market.

We take up a position just inside the market gate. We have twelve banbhs for sale. Eleven of them are a fine sight, long and lean and healthy looking and Daddy has valued them at seven pounds ten shillings per head. Buyers come and haggle but he's not for turning and they're making offers far below his asking price. It's interesting to watch this game of bluff as the surrounding observers join in and take sides with expressions like, "What's between ye"? "Will you cut the difference"? Then there's the spitting on palms and the clasping of hands. But to no avail. He turns down an offer of seven pounds two and six from one cheerful, pleasant jobber and that's the best offer he gets. Many hours later, tired and weary, he lets them go for six pounds fifteen shillings – a drop of seven and six per head and to make matters worse the buyer refuses to include Pigeen in the purchase. Daddy is bitterly disappointed. The deal rankles with him and he can't shake it. It goes round and round in his head like a long-playing record. There would be no treat for a hungry son; I was hoping we would go for meat pies at Greasy Front's Eating House, but instead I have to make do with the packet of Marietta biscuits I had bought earlier. The journey home is tiresome and depressing. The voice in his head gives him no peace and he continues to off-load his regrets on me all the way home. Pigeen lies quietly on the bed of straw beneath us, looking lost and forlorn without his companions. My attempts at levity meet with no success. My philosopher father of the morning's journey is gone and has been replaced by an agitated stranger whose mind is miles away from the present moment.

It was late in the evening when we got home and Mammy had a tasty fry ready for the supper, but nothing could shake his mood. All through supper he kept lamenting his loss while Mammy tried, unsuccessfully, to divert the conversation on to other topics. His regrets lasted for a number of days to follow, causing much unease throughout the house and ending up in a serious quarrel between himself and Mammy. The grand silence set it. Weeks and weeks went by without a word between them. The tension in the house was palpable and the atmosphere depressing. With wounded pride, Daddy played it out to the bitter end. Mammy tried many times to casually ask a question or make a comment, but got no response.

The months rolled by. Pigeen lived alone now in a small shed and was let out regularly on a piece of waste ground to root and forage as nature had intended. When passing by I would stop and talk, more to myself than to him and he would put his snout through the gate, grunting away quietly to himself. I wanted to believe that he still remembered those happy times in our kitchen as a little pet banbh. He was enjoying his free-range, organic existence and was blissfully unaware of the fate that awaited him; something I was keenly aware of and knew was imminent. Killing a pig was something we all had grown up with, but the horror of it never lessened. It was worse on this occasion because Pigeen had once been a much loved pet.

One bright summer afternoon the heavy wooden kitchen table, stripped of its oilcloth covering, is placed in the yard outside the backdoor. A big wooden barrel of scalding water is positioned at one end and a clean galvanised bucket at the other end. Monty McElligott from down the road stands by, sharpening his butcher's knife. With terrible squealing and struggling, poor Pigeen is lifted onto the table on his back with his head protruding outwards and pulled downwards towards the galvanised bucket. Monty makes a deep gash in Pigeen's upturned throat and plunges the sharp knife in the direction of his heart. I watch in horror as his blood flows freely into the bucket. It's over in a matter of minutes. Pigeen's spirit has gone to some happy rooting grounds in the sky and his mortal flesh will provide meat for us for the coming year.

Everybody has their own role to play in the subsequent drama. Sara takes the bucket of blood to the kitchen where she transforms it into the most delicious black pudding using her own secret recipe. We all gather round to watch Pigeen's limp carcass being lowered into the tub of boiling water and left to soak for five minutes or so. It's then hauled back onto the table and the shaving begins. A variety of shaving tools are used; old-style cut-throat razors, sharp knives and even the lids of certain tin boxes that have a sharp edge. 'Many hands make light work' and within a short space of time the carcass is as clean as a new pin. The insides are then removed, almost all of which will be turned into tasty dishes over the coming days. When no one is looking, Monty makes a small cut on the heart and say "Look, that's where I stuck him" and then he cuts off the bladder and shows us how to make a football. He rubs the bladder hard on the table until the skin is elasticised and then inflates it using a goose quill. He ties the neck of the bladder with a piece of twine and tells us it will take a few days to dry out. It's an odd looking football and it feels disgustingly slimy and sticky but we hang it high up on a branch to let it season into leather. By now Mammy and Sara are hard at work, scraping the intestines to remove any fatty bits and then turning them inside out to wash them in the bucket of spring water so as to remove any traces of undigested food. With the guts cleaned and dried Sara begins filling them with the highly flavoured mixture of blood, pinhead oatmeal, onion, finely chopped pork, herbs and seasoning.

To fill the gut Sara uses a Guinness bottle, the end of which has been broken off and smoothed out. The long neck of the bottle is inserted into one end of a length of gut and the mixture poured in using a mug. Mammy squeezes the mixture all along the gut and with a deft twist, at twelve inch intervals, she converts the lot into manageable rings of pudding. These are put into the big black pot on the range and boil for ten minutes. Afterwards they're strung along a straight pole to dry. It's a mouth-watering sight to see this line of black pudding rings. Most will be fried for dinner or supper over the coming weeks, but some will be distributed to friends and neighbours. But a feast awaits us this night. Mammy cooks all the edible offal parts, such as liver, kidneys and the treasured piece of pork steak. Fried with butter and salt, the smell of freshly cooked meat is filling the kitchen and wafting out to the open shed where Daddy has suspended the carcass by the hind legs. He has stretched the carcass open using lengths of freshly-cut ash rods. This will make tomorrow's work easier, but now he finishes up quickly, drawn by the aroma from the kitchen.

Early the following day we lift the flattened out carcass onto the table and, with the precision of a surgeon, Daddy dismembers it into its component parts. Head and limb joints are removed first and then each side is cut up into five rectangular slabs. He stabs the skin all over with a sharp knife and then we begin the slow, tedious task of salting. Each of us are given a slab of meat and for hours we rub the salt in, first on the 'skin side' and then on the 'flesh side' until the big bag of coarse salt is all used up. Grazed fingers begin to smart and sting, but no one leaves their post. We know the salting has to be well done. This meat will have to keep for twelve months or more and then the whole operation will be repeated on another pig. The salted meat is stored in the wooden tub in which Pigeen had taken his farewell bath. After some months in the pickle the slabs are allowed to drip dry and then hung from the hooks on the central wooden beam across our kitchen. Pigeen's dismembered body is back where his life began, turning a dark brown colour as the months go by and casting dark shadows on the ceiling, as the old oil lamp sheds its dim light around our gloomy kitchen.

***

The raising and caring of hens, ducks, geese and turkeys was Mammy's responsibility. These were a valued source of food for the family and provided a welcome treat from our staple diet of bacon and cabbage. Mammy did all the killing. She killed without compunction. The hens were killed by wringing their necks. She would hold the legs high with her left hand and stretch the neck with her right hand. A sudden fluttering of outstretched wings was all we saw and the deed was done. When killing turkeys she would use the handle of the sweeping brush. The poor old turkey's head would be placed under the handle and Mammy would put a foot on both sides and, catching the legs, would pull the turkey upwards until the neck snapped. The killing of geese and ducks was a more gruesome sight because of the blood. Sitting on a chair, with the poor unfortunate bird on her lap, she would bend the beak back against the neck and holding it in that position with her left hand would cut the taut curve of the neck with a sharp knife. While the bird struggled the blood would flow into a saucepan on the ground. Later, the blood would be mixed with herbs, chopped unions and oatmeal and served for supper.

As we grew older, Mammy would give us the job of plucking the duck or the goose. This was best done while the body was still warm. It was slow work and difficult to get all the feathers out especially the downy feathers near the skin. We would try to collect the feathers in a bag but on windy days they blew everywhere. When the job was finished she would cut off the legs and the wings. The geese wings were kept as hand brushes for brushing the dust and ashes around the fire. To get rid of the downy feathers on the skin, Mammy would light a saucerful of methylated spirits and singe the carcass over the flame until all the down was burnt off. On Saturday nights she would prepare the stuffing. Ducks and geese were always stuffed with potato stuffing, while chickens and turkeys were stuffed with bread stuffing. Mammy's Sunday dinners was always worth waiting for; they looked delicious, smelled delicious and tasted delicious.

A dozen or so hens had the run of the small orchard and the goose and gander and their offspring of five or six goslings had the run of the big orchard. The hens were fed with ration and left-over potatoes but the geese were self-sufficient; they lived on grass and, in the autumn, windfall apples. As the goslings grew big and fat their numbers decreased one by one for our Sunday dinners until there was only one left and that would be kept for our Christmas dinner. The goose and gander produced their clutch year after year and this routine was played out until they became too old and they too ended up in the oven. They were never replaced. The white turkey had been introduced to Ireland and became the preferred choice for Christmas dinner, but Daddy always hankered for the taste of roast goose and created a longing in all of us so Mammy sometimes raised ducks and the taste of crispy roast duck was just as good but Mammy would say "there's nothing on a duck but fat".

On one particular summer, however, we had no ducks, except one unique special duck. Somebody had given us a present of a Muscovy duck. It was a beautiful creature with shiny black and white plumage and red streaks on its head and bill. We had no place safe to keep it except with the hens. It was a colourful addition to the small orchard, but little did we realise how xenophobic our hens were. Every day I would go with Mammy to feed the hens to see if Muscovy was alright, but she just stood at a distance watching the hens devour their food. I tried to bring her near but they wouldn't accept her; they wouldn't allow her to eat with them and would peck at her eyes whenever she tried to reach the food. Given time we thought they might accept this beautiful exotic stranger of South American origin, but instead, they pecked her to death.

In springtime it was usual to see two butter-boxes stuffed with hay at the back of the kitchen under the stairs. In these comfortable nests sat two hatching hens, sitting on turkey eggs. In the early days these were the black turkeys that were smaller than their white cousins but were hardier and less prone to sickness and disease. At the end of three weeks hatching, Mammy would take the eggs from the nest and hold them to our ears so that we could hear the tapping sound within the shell. And when the moment finally arrived we watched the shell breaking open to see a little scrawny head struggling to get out. As small children the magic of that moment was fascinating and resonated with some deeper instinct of life. The turkey chicks, though not as pretty as hen chicks, had their own special charm and grew up quickly. One year Mammy bought a bag of special turkey feed in Kelliher's shop in Tralee to feed her fast-growing turkey fledglings and within a few days they all died. So she took the half-used bag back to the shop and told them her story. About a week later a representative from the company who produced the ration called to our house and said they could find nothing wrong with the bag of turkey ration and could not take responsibility for the dead turkeys. The representative had brought some toys and gifts for us, but Daddy intervened and would not accept his offer. He said it was 'blood money', but we didn't understand; we just wanted the shiny new football and other colourful boxes he had brought.

The fox was always a threat so the hens were ushered into their secure abode every night. They spent their days roaming about the small orchard, scratching the ground, as nature intended them to do. In the adjoining big orchard lived the gander and goose with their brood of five goslings. They were fox-proof; it would take a very courageous fox to face down our gander. He was a magnificent specimen, with his shiny white feathers, broad strong wings and long powerful neck at the end of which was a hissing yellow beak. This fellow was not to be trifled with and being within twenty yards of him frightened the living daylights out of me. One evening, as I played in the lower yard, I heard Mammy calling me for supper with the words: "John will you lock in the hens before coming in for supper". I shouted "okay" back and ran to the small orchard where my worst nightmare presented itself.

I stand by the door of the small orchard trying to summon up my courage. The gander, goose and goslings have entered from the big orchard in search of windfall apples. From a safe distance, I begin calling the hens 'tuc, tuc tuc, tuc' many times, but they ignore me. I round up the nearest ones but the geese stand between me and the remaining hens further down the orchard. I walk slowly along by the wall giving them a wide berth but at a certain point the gander spreads out his wings and stretches out his long neck and charges at me with that awful hissing sound. I run back in terror, my heart pounding in my chest. In desperation I throw stones at them from a distance hoping this might drive them back to where they belong in the big orchard, but to no avail. They seem to have my measure; they know I'm scared and are not for moving. The hens, blissfully unaware of my plight, keep foraging away for grubs and worms and refuse to answer my call. Tears of frustration are welling up in my eyes. Time is passing and I'm getting nowhere, but my pride won't allow me to run back to the house and admit failure. Almost an hour has passed and Tom, younger by one year, arrives to find out what's keeping me. I hope he won't sense I was crying as I tell him about the gander. And then, to my amazement, Tom faces down the winged monster. Armed only with a football, he walks straight at him and, as the gander charges, he throws the football, hitting him on his outstretched neck. With loud gaggling the gander retreats and the whole flock go scurrying back towards their own stomping ground. I run after them, throwing the football again to regain my lost confidence and my self-esteem. The gander continues gaggling loudly from a distance; he too has lost face in front of his family and is trying to redeem himself. Trying to sound fearless I shout "Get out and stay out". Tom picks up his football and we retrace our steps. I had learned a valuable lesson: Always carry a football when confronting a gander. As we rounded up the hens and secured them for the night I made up some story that would explain my delay and warned Tom not to tell anyone about my cowardice in the face of the enemy.

***

Towards the end of the year there were two cattle fairs, one in the village of Kilflynn and the other in Abbeydorney. The Abbeydorney fair was known locally as the 'Fair o' the Cross' and was held three weeks and three days before Christmas. It was the last opportunity for farmers, if they were lucky enough to make a sale, to get some money for the festive season. It was a four mile walk in the early hours of a dark winter's morning to get there, but the walk home was far more depressing and disheartening if no sale had been made and this was a common occurrence. The Kilflynn fair was held in early November. It was a time for selling off cattle if fodder was scarce for the winter.

One year Mick drove four two-year old bullocks up the Green Road to Kilflynn and Kevin, Tom and myself were assigned the task of running ahead to block off any exits the cattle might be tempted to try out. We took it in turn to stand at each side-road junction and when Mick and the cattle had gone past we would race ahead to other openings along the way. This made the cattle run also and Mick wasn't able to keep up, so we would have to stop our progress and wait for him. In the chill of the early morning there was steam rising from the backs of the bullocks and a misty cloud blowing from their nostrils. They were all in prime condition after a summer of good grazing. When we got to the village Mick took up a position across from Parker's Pub and we corralled the cattle by the wall that surrounds the graveyard of St. Columba's Protestant Church. It was our job to mind the cattle while Mick dealt with prospective buyers. The buying and selling ritual was an amusing process to observe. We watched Mick's posturing with interest as 'jobbers' approached him. For a while both parties would feign a lack of interest in each other, but eventually after a lot of arguing and hand-slapping a bargain would be struck. Mick was good at this game and even though he hadn't got the price he was looking for he seemed satisfied with the deal. The bullocks were driven away and we were free to wander about. Mick had a greeting for everyone as we made our way towards the pub. We drank red lemonade while he drank numerous pints of Guinness and talked with fellow cattle drovers about the depressed state of cattle prices.

On the way home, a sixth class girl walked with us as far as the school cross. She wore a tight-fitting beige coat with a brown collar which was partially covered by her blonde, shoulder-length hair. Her high leather boots had shiny brass buckles on the instep that matched the buckle on her belt. She looked quite the picture of a young lady; so very different from the girl I met up with in school every day. I was seeing her in a new light. She was trying to make conversation with me but I had suddenly become acutely aware of my torn jumper, my old worn trousers and muddy boots and got totally tongue-tied. I wanted to say something that would impress her but couldn't string two sentences together. I was relieved when our directions parted at the school cross and we headed down the Green Road. Mick was in great humour. He had sold the cattle at a fairly good price and had drank more pints than he usually would. He had been having prostate problems for some time and the heavy intake of Guinness didn't help. So he had to stop many times to pee and each time he would say: "Now boys, it's time to pee" and we obliged every time even though we were all peed out. He made us laugh, telling us funny ridiculous yarns. We enjoyed the walk, the laughter and that precious time together. Mick had been a second father to us all our lives and we knew he cared about us. Little did we know then that within a few years the uncle we knew and loved so much would be lost to us forever and be replaced by a total stranger that didn't seem to know us.

The town of Listowel held regular cattle fairs throughout the year, after which the town square and footpaths would be covered in cow dung. Shop owners and publicans would be seen with brushes and buckets of water cleaning their shop-fronts. It was a price they were willing to pay for the increased trade the fairs brought to the town. Taking cattle to Listowel fairs meant hiring a tractor and trailer. Pats Egan, if he was available, was always Daddy's first choice because he was very reliable and very careful. Daddy had made a special loading bay behind the turf house. This was where the main chimney block of the Great House had stood. The chimney stack was a massive tower block reaching forty feet into the air and, because it was difficult and dangerous to demolish, it had remained standing long after most of the house had been knocked down. But Thady Buckley needed stones to build a new house and in a few weeks had brought it down almost to ground level. A four foot high base was left and this became the ramp for loading cattle. The cattle would be driven in to the rectangular enclosure, once the main hall of the old house, and corralled behind the loading ramp. Three-year old bullocks, used to roaming freely in the fields didn't like being penned up like this, so a barricade would be erected on either side of the trailer to prevent them from trying to escape. After much jostling they would finally be loaded and Pats and Daddy would set off on the long road to Listowel, a road that Kitchener, as a teenager, had walked with his father's cattle one hundred years earlier.

Chapter Six

Field Work

Daddy and Mick were a good team but they had clearly defined roles. Mick was the elder brother and had very definite ideas and this had a bearing on their relationship. Daddy would often defer to Mick in making decisions about farm work. Certain jobs would always be left to him, like operating the hay slinger in the meadow. The slinger was a contraption for gathering up the 'saved' hay. It was made of seven long wooden prongs sticking forward from a heavy plank. This was pulled by a single horse using very long chains. When the prongs had gathered up an enormous bundle of hay Mick, with great skill, would topple it head over heels beside where a wynd (stack) of hay was being made. This was demanding work and on hot sunny days he would take off his jumper and tie a piece of binder twine around his waist as a belt. The big heavy braces that normally held up his trousers would be removed from off his shoulders to hang loosely on either side. The long sleeves of his collarless shirt would be rolled up to the elbows exposing pale grey-haired arms. He would daringly open the top buttons of the shirt to chest level revealing a patch of very white skin that contrasted vividly with the tanned skin of his neck. The peaked cap that was perched precariously on his head would be lifted occasionally to wipe the sweat from his forehead.

Daddy, on the other hand, would remain fully covered from head to toe with an additional sun shield added. He would place a handkerchief under his peaked cap and stretch it over his ears and the back of his neck like a desert trooper. The tops of his ears were particularly sensitive to the sun and Mammy would skim the cream off the milk jug to soothe his burnt ears. I loved those warm sunny days in the meadow but often, at the close of day, endured the agony of sunburnt arms and legs. She would soothe my burning skin with a paste she made from bread soda and milk. It helped a little but rubbing against the bed covers at night kept me awake. On other meadow days when rain was threatening and much work remaining Mick would work at a frantic pace. In a desperate race against time he would attempt to pile one slinger load on top of another so that the high mound of hay could be quickly topped off and tied down. So many heartbreaking days in the meadow would start out gloriously sunny and change to rain in the afternoon just as the hay was ready for wynding. On one such day, we were saving hay in the 'small dale' when thunder clouds rolled in and the rain poured down. In the midst of his frustration Daddy showed us how to make hay-umbrellas with our pikes and we walked home through the rain and the lightning without getting wet. But sometimes, when God smiled down, a week of glorious sunny days would arrive at the right time and working in the meadow became a most enjoyable experience.

When the morning sun had dried off the dew the swathes of hay would be turned to allow the underneath to dry out. After a few hours it would be ready to be made into wynds. Three swathes would then be raked into one and Mick would begin the slinging. Gradually the wynds would mushroom up here and there all over the meadow; each one tied down with the yellow binder-twine. They would then be 'pulled from' tightly all around the base and raked down smoothly giving them the shape of giant eggs. Constructing a wynd of hay was a tricky business and a lot depended on the person standing on top of it while it was being made. The construction required a great deal of skill, spreading the hay evenly all around and tapering in as it neared the top. Holding one's balance was difficult when the wynd wobbled, and timing the upcoming thrust of the three-pronged pike, as you grabbed the hay, was essential for safety. One day in the front meadow, Daddy decided to stand on the wynd to show us how to do it properly, but the sharp steel prong of an up-thrust pike went through his middle finger bursting off the nail. His clothes and shoes were covered in blood as he made his way into the house. Mammy quickly reached for the biscuit tin that served as our first-aid kit and dressed the wound, wrapping it tightly in a white bandage. He was back in the meadow within the hour and continued working all day as his bandaged finger took on a deep red colour.

'Many hands make light work' and a *meitheal* (group) of workers anywhere creates its own dynamic and generates a new energy that makes the work easier and more exciting. Sometimes, neighbouring men would appear out of nowhere with their hay pikes on their shoulders and 'fall in' with the work. No explanation was needed; this was how the meitheal system operated before the mechanisation of farming. It had been part of the culture of Irish farm work for generations. It was an accepted code; a day's labour given would be returned in some form or other – a sort of credit system that allowed for the borrowing of a horse or farm machine or a helping hand when needed. It afforded a sense of community and connectedness that added much to our quality of life and the simple enjoyment of country living.

On one glorious afternoon we were wynding hay in the Barrack field when out of the blue Tomásín Whelan appeared; like the yank in Glenanaar, 'gorgeous and caparisoned from head to toe in all kinds of sartorial splendour'. His dark suit and hat contrasted starkly with his gleaming white shirt. He was home from England on his yearly holiday to see his family; a life-pattern common to many emigrants in rural Ireland at that time. Tomásín had been a close buddy of Daddy and Mick growing up. Even as a young child I could sense the warmth with which they greeted him. Their conversation was more laughter than talk; Mick 'having a go at him' about this and that. Mick was well able 'to dish it out' back then, but it was all in good fun. I often wondered if Mick envied Tomásín for taking the emigrant ship and carving out his own life. Mick had chosen a safer, less adventurous path.

True to *meitheal* form, Tomásín whipped off his coat and placed it neatly at the side of the meadow. His hat and waistcoat he kept on and without further ado he began piking the hay. He was a good worker but a regular guffaw of laughter would erupt from his side of the wynd that brought everything to a stop. It became infectious; he had us all laughing, young and old alike. A shout went up as Mammy was spotted at the meadow gate carrying two heavy-looking bags. "Taytime, thank God"! Pikes were stuck in the ground and everybody sought out a comfortable spot to sit down or lie down and wait for the large bottles of tea and the thick slices of bread to be handed around. There would also be a few 'goodies'; Mammy was good at baking and would have done fresh scones and 'curny cake'. The banter continued; she was very fond of Tomásín and invited him for supper that night. She set the table in the parlour using her best china, a very colourful and delicate set that only saw the light of day on these special occasions. The fare was rustic but wholesome; lettuce leaves freshly picked from her own little garden patch and a plateful of ham and hard boiled eggs, with plenty of brown bread and fresh scones.

All the adults were seated around the table but we as children were not allowed in. We could hear what was going on in the parlour as we ate our own supper in the kitchen. The merriment and revelry continued for hours; conversation interspersed with jokes and stories and bouts of laughter; all on a few pots of tea. Mammy had a loud laugh and Tomásín knew how 'to press her buttons'. She would be weak from laughter by the end of the night with tears rolling down her face. Daddy would add to the banter and if the scones had only a few currants in them he would say, "You must have been standing at the door when you flung the currants in". Like Tomásín, Daddy also had a skill in telling jokes. He had a way of under-telling the story that made the punch line all the more effective and would throw in many witticisms at the right time to keep the laughter going. Down through the years Tomásín drifted in and out of their lives in this way, for brief exhilarating re-unions.

The meadows were cut one at a time to offset the risk of a prolonged spell of wet weather. Daddy did the cutting using a mowing machine drawn by the two horses. The machine had a long pole stretching out in front with a quoin bolted on at the end that could be hooked up to the horses' collars. The chains for pulling the machine were hooked on to the hames that was clasped around the collar. Two long ropes for guiding the horses reached back to where Daddy was sitting on the machine. The cutting apparatus was ingenious. On entering the meadow it was lowered to a horizontal position, a few inches above the ground, supported by a small wheel. A long metal strip with twelve triangular blades, riveted firmly in place, was slotted in through the mowing bar and these moved back and forth through the prongs when the machine was in motion. If the blades were well sharpened they gave a neat cut so Mick would sit at the headland sharpening the replacement blade sections, which had to be changed regularly.

One sunny summer's morning, with the smell of the newly mown hay wafting in the air, I follow behind the mowing machine, hypnotised by its staccato rattling noise. Daddy and the horses are moving at a steady pace as they turn a field of tall grass into an artistic geometric design, each swathe laid out perfectly. He knows the movement of the horses well and they respond readily to his commands. As he turns the machine at the corner he sees me behind him with something in my hand. "What have you got there, Kesteral", he shouts over the noise of the mowing machine. 'Kesteral' is an affectionate term he uses for all of us from time to time, the origins of which remain a mystery. I show him the legless corncrake chick that's flapping about in my hand. Unable to run fast the poor little thing has been butchered by the blade, its legs severed just below the knee joint. Daddy gets down off the machine, takes the bloodied corncrake from my hand and tells me to look away as he crushes it underfoot. Its agony is over in an instant. The horses are restless, swishing their long tails to combat the hordes of horse-flies that always follow the machine. He gives me a sympathetic grin and quickly gets back to work. A pungent smell of fresh horse dung hangs in the air as I follow along behind. When I come near the gate I decide to leave; I don't want to come across any more amputee corncrakes.

After a few days the swathes of hay were turned over using wooden hand-rakes or hay-pikes, or in our case as children using our feet. The long stalks were so entwined that when you kicked up the heavy end it brought the rest with it. As children we could keep kicking it up and over at ease, moving so fast that the adults wouldn't be able to keep up the pace. On one particular day while doing this work I had a terrible toothache and the day seemed endless. The only relief I got was from sipping cold water, but the relief only lasted a few minutes, so I spent the whole day, in agony, sipping cold water. Luckily, the meadow was near the Promsey well, where I could refill the bottle. The following morning a dentist in Listowel removed the offending tooth and I was so relieved to be pain free that I gladly joined them in the meadow in the afternoon. Each year the same five meadows provided enough hay for the winter, if weather conditions allowed it to be properly saved. The wynds, if well made, were weatherproof and would be let sit in the meadows for a number of weeks to season.

Not only did we have to help save our own hay, we had to help save Uncle Johnny's hay as well. He was the eldest of the Galvin family and as tradition would have it, he inherited the family farm at Ballyrehan. This large farm had been the Galvin homestead for generations. The story handed down to us was that four Galvin brothers had come from a place called Ballyoneen to work as tenant farmers on Ponsonby land towards the end of the eighteenth century. They each took up tenancy of about fifty acres. For reasons unknown one brother emigrated to America and two of the farms became one and was farmed by my great grandfather. In turn my grandfather inherited the farm even though he was a student at Maynooth. Why a clerical student became a farmer we can only surmise but this became his life and he was known to all and sundry as 'The Saggart'. In turn, Uncle Johnny inherited the farm. Being the heir to the family home meant putting wedding prospects on hold until 'the nest was empty', so when he married Kathy Dillane from Listowel, they were both of mature years and had no offspring to help with the farm work. So Kevin, Tom and I were often 'volunteered' into helping him out.

It was clear to us, even from a young age, that Johnny had lost interest in his farm and was only going through the motions at a subsistence level.

The once fine buildings in his farmyard were now in a dilapidated state, with mountains of dung outside each cowshed. The dung and straw in the calf sheds hadn't been removed in decades so that his few confused calves were living 'upstairs' with their heads touching the wooden beams of the roof. Many of his calves died every year and were left there to rot until the stink became unbearable. Entering his cowshed to hand-milk the cows meant wading through a pond of urine, dung and water and this he and Kathy endured, every morning and evening. The traditional style, single-storey, long thatched house, which had been the Galvin homestead was slowly collapsing all round them. The thatched roof over one of the bedrooms had fallen in and smothered the bed and the furniture and was left in that state. The whole scene was totally depressing. Kathy rarely left the precincts of the house and farmyard. Most of her life was spent in isolation and loneliness. Johnny's daily trips to Lixnaw creamery with his small tank of milk was her only lifeline to the outside world. He would stop off for a few pints on the way home and she would eagerly await his reports of local gossip and when he delivered the 'Kerryman' on Fridays she would pour over every page for every last morsel of news.

Every summer we would help him save his hay. Having worked hard all day in the meadow we would have expectations of some small financial reward in the late evening but mostly all we ever got was, "God spare ye the health, lads. Will I see ye tomorrow?" On one particular morrow a well attired figure approached from the distant corner of the meadow, carrying a two pronged pike on his shoulder. Our hearts jumped with joy when we recognised that mystery man from London, Tomásín Whelan. We couldn't understand why he should choose to spend his holiday time in this way; piking hay in Johnny Galvin's meadow on a sweaty afternoon. But he had grown up as a neighbour of the Galvin family long years before and this seemed to be his way of connecting with the past. We were glad of his presence because we knew we were in for a day of laughter and joking. When a prong on the wooden slinger came loose Johnny said he'd have to go for his tool box. Tomásín, knowing Johnny's form well, described the tool box as "A few rusty nails and a stone" but nevertheless the slinger got fixed and by the end of the day the whole meadow was in wynds and neatly raked. Tomásín summed up our day in his parting comment, "A good day's work for three beginners and two auld latter ends". Weeks later I would visualise him in my mind's eye sitting in a bar somewhere in London, telling tales of other days or passing time in a lonely bedsit, thinking of Johnny Galvin's meadow and the three *garsúns* (young boys) who laughed at his jokes.

At the end of that day, Kathy was waiting for us by the gate with three shiny half-crowns. It almost made it all worthwhile as she tried, in her own peculiar way, to express her thanks. We felt that Kathy was always a bit distant towards us; perhaps fearing we had designs on her precious farm, or maybe dreaming regretful dreams of what might have been. But Johnny was always jovial and pleasant and his indomitable spirit never wavered even when, as an old man, he had suffered a stroke. He continued to walk down the road, wearing his long grey overcoat and matching grey hat. Those who asked him how he was all got the same grim answer: "I'll bury a lot of 'em yet" and so he did. By then the whole roof had collapsed in the thatched house and he and Kathy had moved into the house in the Barrack field that had been built for Sara and Lizzie. After Kathy's death Mammy became his carer. She cooked his meals, did his washing and occasionally supplied him with a bottle of whiskey. Every evening after his supper he would ask her to light his pipe. This meant finely cutting the plug tobacco, kneading it between her palms and then trying to light his old congealed pipe, as the foul taste of tobacco tar seeped into her mouth. Half a box of matches later she would hand him his smoking pipe and just as she was about to make her way home she would hear him shout; "Come back, 'tis gone out again!".

For years Johnny used to give Daddy the use of the 'mashes' for hay. These two fields, close to the river Brick, were called the 'mashes' (a corruption of marshes) as they were subject to flooding. The hay was of poor quality but it was better than no fodder if spring came late. Bringing the hay from the 'mashes' to the hayshed at Crotta was a slow and tedious process. It was a long haul up the main road. This was Mick's responsibility and patiently he would go up and down that road with the hay-car, making four or five trips each day for many weeks. The hay-car, sometimes called a hay-float, was a low-level horse drawn car with a broad surface of smooth boards on which a single wynd of hay would sit. I loved to accompany Mick on these trips. It was the easier option. While Kevin and Tom were straining every muscle piking hay into the shed I was enjoying the fresh air and the open road. When passing Stack's house, their three children, Kathleen, Bobs and Frances, would steal a ride on the back of the hay car. Mick would shout at them to keep away and would pull Moll to a halt and make them get off. They would wave at us as we moved on and would be waiting to sneak a ride as we returned with the wynd of hay.

On reaching the meadow, Mick gives me the reins to make me feel more grown up while he sits there on the other side crooning his favourite song about the young man who had been shot by the Black and Tans. He comes to a strong slow finish with the line: 'It's underneath this deep green sod young D. J. Allman lies'. He sits there quietly, all sung out. The mashes are now covered in a deep green coating of lush after-grass and I hate to see the wheels of the hay-car and Moll's hooves flatten it to the ground. I want to go around by the verge to avoid trampling the pristine carpet of grass but Mick insists I go straight to the nearest wynd. He reverses the hay-car into position. Together we drag the heavy ropes around the wynd, tucking them underneath all round and clasping the iron hooks together at the back. The ropes are attached to two large pulley wheels at the front and with the use of a cogged wheel and a winding handle, the wynd of hay is slowly inched forward onto the sloping hay-car.

This done I climb up to the top of the wynd and lie there, face downwards, for the homeward journey. The sturdy Moll looks much smaller from this viewpoint as she strains against the collar with Mick urging her onwards with a gentle flick of the reins. High up in my comfortable bed I drift into dream world, lulled by the rhythmic movement of the hay-car. A warm breeze caresses my face as I gaze at the broad patchwork of surrounding countryside, fields of all shapes and sizes, guarded by their boundaries of green hedges and tall trees. A flock of crows are winging their way towards the tall trees and their discordant screech is grating on my ears and I'm wondering why God didn't give them a sweet singing voice like other birds. Cows are grazing contentedly in the lush after-grass in Stack's meadow. Some of them have filled their bellies and have laid down to chew the cud, while they wait to be rounded up for milking. Away in the distance the clear silhouette of Stack's Mountain stands out against a deep blue sky. As far as my eye can see it stretches; a long low ridge of purple and brown. The mountain carries the name of the deStack family since the Norman conquest of Kerry in the thirteenth century. They were once the masters of all they surveyed. Cromwell changed all that but the family name of Stack lives on in the farm by which we are now passing. The road is quiet except for an occasional black car or a grey TVO tractor. Jerry Nolan is leaning against the pillar of his front gate, I wave at him as we saunter along and he waves back. The clip-clop of Moll's hooves and the crunching of the iron wheels on the road are the only sounds to be heard. Everything seems perfect. I'm feeling very relaxed on top of the wynd and my eyes are beginning to close, but a shock awakening brings me to my senses.

As we passed by Stack's house little Frances came out from her hiding place. She hopped on to the back of the hay-car for a short ride and then hopped off again and, without looking, dashed towards her front garden. A horrible screeching of brakes shattered my peaceful dreams. Mick reined in Moll with a jerk, tied her to a bush and ran to investigate. His worst fears were realised. There was Frances twisted up underneath the front axle of the car. Her wailing voice was filling the air. Lil, her mother ran to the scene in a panic, followed by Mike Shea who worked in the farmyard. The screaming from underneath the car grew louder as Mick and Mike Shea and the driver lifted the front of the car and her mother gently eased her out. Car grease and blood covered her face, her hair and her clothes. Lil was holding the little battered body in her arms as the driver turned the car around and rushed them to Tralee hospital. Mick was totally devastated and Mike Shea had tears in his eyes because Frances was his little princess. When Mick got home he couldn't do any more work. He was drawn and pale and listless. He felt somehow responsible for what had happened and he feared the worst. He took it very badly and said he couldn't bring himself to ever again bring hay from the 'mashes'. Daddy would have to complete the work, but not until many days later when he heard the news that Frances was going to be okay.

Frances overcame her ordeal and in time reverted to her dare-devil activities. She grew up to be a beautiful girl who seemed none the worse for her awful experience. Some days later Matt Stack, Frances' father, came to visit Mick to re-assure him that everything would be fine. The Stack family had been very close friends down through the years. Matt was held in very high regard by everyone in the neighbourhood; always good-humoured and willing to help whenever needed. Mammy was particularly fond of him and always sought out his company. It was little wonder then that the news of his sudden death, about a year later, caused her to drive the car against a ditch on the Green Road. It was on a rainy morning driving us to school that she stopped to give Teresa Crowley a lift. As the car was taking off Teresa blurted out "Matt Stack is dead, he was killed in a car crash last night". With the shock of hearing these words Mammy looked back and lost concentration and drove the car into the ditch. She sat there for a long time in a state of total disbelief, repeating the same question, "Are you sure, Teresa?" The whole neighbourhood experienced the same shock. His tragic death cast a dark shadow over the local community for many months. There would be no more hurling in Stack's field, which was situated across the road from their house. But life went on. After a few weeks Crotta hurling club got the use of another field in a neighbouring farm but for various reasons it didn't work out and eventually, with the healing of time, the hurling came back to Stack's field again.

Chapter Six

***

During school term we usually arrived home at 3.30 and Sara would have our dinner ready. This consisted of whatever was left from the adults' dinner and hardly ever varied; bacon, cabbage and boiled potatoes with a mug of milk. I had a poor appetite so she would sometimes mash the potatoes into pandy and shape it like a bar of chocolate to entice me to eat it. After eating we were expected to help out with whatever farm work was in progress in the meadows, or the garden. Planting potatoes was the job I hated most. All the preparation work would be completed by Daddy and Mick. Ploughing took place in early spring and different fields would be ploughed in rotation using a single board plough pulled by the two horses. Later the soil got prepared for sowing using the spring harrow or the common harrow. Usually we had two tilled fields; one for barley or oats, the other for potatoes, turnips and other vegetables. The drills would be opened for the potatoes using a double-board plough called a ridger. Seed potatoes were planted by hand, twelve inches apart, placed onto the wet sticky farmyard manure that had been spread along the furrow. As children we were all given our bucketful of seed potatoes and had to trudge our way painfully along the manure-covered furrows from end to end of the garden. When we were finished planting, the ridger was used to close the furrows, making straight A-shaped drills the length of the garden. It was a pretty sight to behold at the end of the day; an artistic masterpiece of brown clay speckled with grey stones. Within a few months nature would perform her miracle and each potato seed would produce ten or twelve beautiful large tubers. Kerr's Pink and Sharpe's Express were the favourites but Golden Wonders were good too because they stored well through the winter. Turnip planting, on the other hand, was easy. Turnip seeds were planted with a machine. I was fascinated by this machine. It was an ingenious piece of equipment and could be pulled easily by one horse, depositing the tiny seeds in a straight line on the top of the drill. In a few days, parallel rows of green shoots would appear on the dark-brown earth. But the weeds were always 'waiting in the wings' and would take over very quickly if we weren't ready with hoes to "redden the furrow", as Mick used to say.

Within a few weeks the turnips were ready for 'thinning'. From a young age we had to play our part and Daddy pushed hard. He had an expression which he would direct at us to counteract our idle chatter. It went: "Grady's men, if you talk to them they do nothing". We never knew who Grady was but obviously Daddy was cast in the same mould. The thinning of turnips was gruelling work. We would wrap jute bags around our legs and secure them with bits of twine above and below the knee so that we looked like the cowboys branding steers in our comic books. Down on all fours, we crawled together the full length of the garden, each along his own drill, pulling out the weeds and excess plants and leaving the strongest plants growing about six inches apart. During dry spells the drills became hard and crusted which often resulted in broken fingernails and cut knuckles. When I was down on all fours the long drills seemed endless. It was best not to look up; to focus on the spot directly in front was the only way of getting through it. I was meticulous about plucking out the weeds by their roots and removing any stones that might hinder growth. Because of this I would fall behind the others as they moved at a faster pace along the drills but Daddy, seeing my plight, was able to do bits of my drill as well as his own, to help me catch up. As we neared the headland Mick would take off at full speed and somebody would quip, "*Make Way for Lord Kitchener*" and Mick would finish well ahead of the rest of us. He would never be beaten at thinning and would race to the end regardless of the quality of his work. His need to be the best and fastest was strong within him.

By mid-summer the potato garden would be covered in lush green potato stalks and Maurice Neenan would be summoned to do the spraying. He had a horse-drawn spraying machine which would move up and down the garden covering five drills at a time. A dilution of bluestone and washing soda was sprayed onto the stalks to prevent the dreaded blight. It was over a hundred years since the Great Famine but the horror of that disaster had left a fearful mark on folk memory and it sometimes entered the conversation around our dinner table. Under what circumstances the Galvin household at Ballyrehan survived the famine is not known, but presumably they weren't totally dependent on the potato.

Over one million people starved to death in the countryside and of the million that emigrated many died on board the 'coffin ships' across the Atlantic. With the wisdom of hindsight historians inform us that the famine was a disaster waiting to happen. In the early decades of the nineteenth century, one variety of potato, the *Irish Lumper*, had become the staple food for the poor. Subdivision of land had resulted in individual holdings being very small and no other crop would suffice to feed the large families of the time. The lack of genetic variability in this single crop meant that *phytophthora infestans*, commonly known as the *blight*, would wipe out the entire potato crop. This disease, believed to have been brought into the country on the merchant ships, brought down the landlord classes, including the Ponsonbys of Crotta. Without the rent from the tenant farmers they faced bankruptcy and were forced to sell their estates. The humble spud had played a major role in shaping Irish history, but Maurice Neenan's spraying machine and a small quantity of bluestone and washing soda guaranteed our potato crop each year.

Potatoes were our staple food and needed to be carefully stored and preserved to ensure they lasted until the following year's crop was ready. But rats were a problem. When the potatoes were stored in the shed they attracted rats so Daddy used to set rat-traps in strategic locations. These were old style metal traps with a very strong spring which caught the rat by the legs if he happened to stand on the metal plate. Seeing a live rat squealing in the trap was a horrific sight but it never seemed to bother Daddy. Holding the trap in his hand he would walk to an open field with the terrier following behind. He would release the rat so that the terrier could do what terriers were programmed to do. It wasn't a fair contest but it was both fascinating and horrifying to see this evolutionary ritual being played out. I watched from a distance as the terrier grabbed the rat by the back and shook him fiercely until he was dead.

In a corner of this field there's a large clump of briars and as I walk by I notice a little tunnel-shaped track running in through the tall grass. It looks like a rat-run so I secretly take a rat trap from the shed and set it there. The following day I come to inspect the trap and find a blackbird caught by both legs. It's still alive though weak from pain and fluttering. Feelings of guilt engulf me and my heart feels sad for the poor little creature. This beautiful bird had suffered all night because of me. I know it has to be killed to end its pain but I have never taken the life of a living creature, except flies in the kitchen. Mick would have to do it for me. I search everywhere but cannot find him, so I go back to the scene of the crime. I release the blackbird from the trap and it flutters its wings in a feeble attempt to escape. I take the rat-trap back to the shed, pick up an old shovel and return. The helpless bird is lying there unable to move but its eyes are open, looking at me about to commit murder. I close my eyes and swing the shovel. The blackbird's pain is over but the trauma of killing it stays with me for many weeks.

The field that lay to the west of the orchard wall was always full of rabbits. They ventured out late in the evening to eat grass. It was a thrilling experience to sneak quietly to the end of the wall and watch them nibbling away to their hearts' content. One evening as I peered around the corner there was a very young rabbit within ten feet of where I was standing. I was facing into the sun as it shone upon him and he didn't see me. Like a stalking tiger I moved forward slowly and noiselessly. The small rabbit was totally absorbed in what he was doing. My hands were stretched out in front in a grabbing position and my heart was pounding wildly. I was so close and he hadn't sensed my presence. I held my breath and dived. I had him. He was a little beauty and he didn't really seem scared. I cradled him in my left hand and began to rub his soft fur, but in an instant he leaped out of my arms and scampered into a burrow. My heart was still pounding as I gazed around. All the other rabbits had disappeared except for two and they were dead. Daddy had set snares to catch them. I brought them home. Tomorrow he would skin them and we would have rabbit stew for dinner.

Rabbit stew was a welcome change from bacon and cabbage and was very tasty. Mammy made a delicious soup that Daddy liked to sip from a mug. Rabbits had been a source of food in Ireland since they were introduced into the country by the Normans in the thirteenth century. They had become very plentiful in all parts of Ireland but by the mid fifties their days were numbered. Farmers were complaining that the rabbits were eating all their crops and the Government introduced a contagious viral disease into the rabbit population called myxomatosis. It was a slow lingering, debilitating disease that gave them a horrible death. They could be seen staggering blindly in the fields and along the roads, their eyes bulging out of their sockets. Dogs and cats avoided them like the plague. Within a few years they were wiped out. Their gracious presence was obliterated from the landscape and a crucial link in the food chain had been removed. But now the farmers could look forward to bigger yields from their labour.

***

Interchanged with work in the gardens and meadows was work in the bog. Turf was our source of fuel; we never burned coal and rarely cut logs. We owned a long bank of bog in Ballinagar and we cut the turf in the traditional way. The traditional method of *sleán* (turf-spade) cut turf required a *meitheal* of workers. To get the job done relatively quickly two *sleán* teams, of at least four men each, were required. The first job was to remove the grassy sod along the strip of bank to be cut. In our bog the bank of turf was five sods deep with the brown turf at the top and the black pasty sods coming from the oldest layers deeper down. The cut of the *sleáns* exposed millennia of archival history of bog formation dating back to post ice-age Ireland and occasionally a tree branch thousands of years old would be unearthed. A brief moment of speculation might be spent marvelling at such a find and this would be laced with jokes and witticisms before returning to the serious business of turf-cutting. By the end of the day a layer of dark brown sods covered the purple heather along the bank.

There was no real role for children in the bog during the turf-cutting days, except to have fun collecting bog cotton and searching for lark's nests. The mother lark would play her usual trick of pretending to have a broken wing to lure us after her and away from the nest. But we were wise to this trick and would find her little cosy nest hidden deep in the heather. It was funny to observe the four or five little scrawny fledglings opening their mouths in expectation of food. We would carefully close the heather over the nest and retrace our steps to the campfire. Our job was to keep an eye on the fire while the kettle was boiling and have it ready when the men arrived for lunch. Lunch in the bog was a most enjoyable experience. The pleasant smell of the turf smoke filled the air as the *meitheal* of men took up their positions around the fire. No tea pot was used; the tea-leaves were added to the large kettle and left to stew for a while. Big mugs of sweet tea washed down Mammy's brown bread and the roasted chickens she had prepared were torn apart by hand and eaten without decorum. Hard boiled eggs and potato salad added to the feast and to finish there was sweet cake and biscuits. Then came the well-earned rest when men would stretch full length on the soft boggy ground and smoke cigarettes or have a short snooze.

The turning and stooking of turf was a different matter. This required many hands, big and small, and so we experienced the back-breaking labour of saving turf. On the surface turning turf looks easy. You simply turn each sod back into the path of the previous sod, thereby exposing the wet underside to the air and sun. But in doing this work you had to remain in a crouched position so that by the end of the day your back felt like it was about to break. Stooking was not so bad except on windy days you might get caught on the windward side of the stook and get turf dust in your eyes. I have painful memories of one such day when the turf dust got in my eyes; a day of excruciating agony and slow torture with no reprieve until six o'clock that evening. When I got home Mammy saw the state of my inflamed eyelids and got working on them with an Optrex wash and a piece of gauze and with great difficulty eased out the stubborn turf-dust from both my eyes. My eyes were painfully sore and looked red and raw. At the supper table she gave out to Daddy saying that a child should not be let suffer like that all day but he passed it off with an attempted joke. He said my eyes looked like they had been hemmed with red thread. She was not amused and I was at such a low ebb that I started to cry, but like so many other misadventures I accepted it all as the way of the world and hoped for better days to come.

***

As summer blended into autumn the field of barley would become a field of swaying gold. But there were many summers when it didn't; instead it lodged. Nothing depressed Daddy so much as seeing large patches of corn lying flat on the ground because of heavy rain and the crows. Huge flocks of crows would land on the slightly bent stalks and trample them flat into the ground. This would cause the ripened grains to sprout and spoil the crop. Scarecrows didn't work very well so Daddy would make a little shelter in the middle of the field and the poor sheepdog would have to stand guard there until the corn was ready for cutting. He would secure the dog with a chain and because it was lonely and away from home it would keep on barking constantly. Whenever I brought food to him I would stay beside him for a long time to ease his loneliness and be rewarded with continuous licking of my hands. It was difficult to say goodbye and leave him there all alone. Sometimes a cat was used for the same purpose by keeping it caged in the field. But in spite of the barking and the lonesome cry of the cat, the crows would still do their handiwork.

When the corn was ready for cutting, John McCarthy who owned a 'Reaper and Binder', would be contacted. This machine was a wonderful invention powered by a tractor. It would cut the corn, bind it tightly into sheaves and drop the bound sheaves in straight rows for us to build into cone-shaped stooks. But before the Reaper and Binder could do its work the field had to be prepared. A few days beforehand Mick would take down the scythe from its safe place on the shed wall and begin to sharpen the blade with a whet-stone he had bought at the creamery store that morning. The scythe had been the traditional cutting tool for generations. It was a dangerous weapon, with a long curved blade and a long crooked handle, but in Mick's hands it was an artist's tool; it was poetry in motion as he cut the lodged corn. He would cut out all the lodged patches and also clear a pathway around the edge of the field. We followed along behind him binding the sheaves in the traditional way using a twist of nine or ten stalks. Preparing the field in this way made it easier for John McCarthy's Reaper and Binder and none of the crop was lost. At the end of the day when we had stooked all the sheaves we would stop at the gate and look back at our work. It was a truly satisfying sight to see all those straight rows of stooks bathed in the golden light of the evening sun. I would gaze in wonder at the beauty of the scene - the smoothly cut stubble field stretching off into the distance like a vast expanse of tawny carpet. In a few weeks the precious sheaves of corn would be brought into the farmyard for threshing.

It was late autumn and the sheaves of barley had been built into two giant ricks, standing eight feet apart, in the farmyard. Dick Shanahan's huge threshing machine, pulled by an old Fordson Major tractor, had rumbled into our yard the evening before and had been parked strategically between the two ricks. The following morning a *meitheal* had gathered for the big event and as they waited for work to start they were greatly amused by the sight of Dan and his cat. When Dan was four he had an inseparable friend, a big grey tomcat. This cat had a damaged eye as if he were winking at passers-by. I had damaged his eye when I hit him with a stick one day, because he was on top of my favourite black cat, biting her neck; at least that's what I thought he was doing. How Dan trained the cat to do 'piggyback' nobody knew but he would walk around with the cat on his back, just like a long furry schoolbag. The cat's front paws stuck out over Dan's shoulders and he held them with his hands while the cat gazed around contentedly from this uncomfortable position. The men were joking with Dan but he was far too interested in the machine and how it worked to take any notice of their comments.

A long heavy belt was connected to the 'thrasher' from a spinning wheel at the side of the tractor. The tractor engine was started and a loud roaring sound meant the threshing mechanism was in motion. We stood and watched as the men took up their stations. There was a man on top of each rick piking sheaves to the two men at the drum's mouth on top of the thrasher. One man cut the binding twine on the sheaf and the other fed it into the drum, ears downward. Each time a sheaf was fed in the roar of the drum grew louder and conversation had to be carried on in loud voices. This was dangerous work and stories were told of hands being mangled and fingers shredded. Three men were working at the rear end of the thrasher where the loose straw came out. This was taken away and built into a giant stack further up the yard. At the other end four chutes funnelled the grain into large jute bags. Daddy was supervising this operation in case any grain got spilled. When the bags were full they had to be carried across the yard and up a flight of stairs into the loft above the stables of the old Great House. Hay was stacked at one side of the loft but at the other side the three-hundred-year-old floorboards were still perfectly sound and here the grain was spread out to dry. The old wooden stairs had seen better days but Daddy had done a rough and ready repair job. Three men with strong backs took on the job of hauling the bags of grain, Matt

Stack, James McCarthy and John Joe Whelan. Dick Whittington and his cat went with them and was a source of great amusement in the midst of their back-breaking work. Even though he was coming in their way and slowing down their work they patiently allowed him to climb the stairs in front of them and each of them in turn would give him a piggyback ride on the way back down, with the cat still on his back.

In the midst of all the activity and commotion Mick walked around with a bucketful of porter and a mug from the kitchen. It was sweaty, dusty work and one needed something with a kick to wash down the dust and chaff gathered in the mouth. The same large mug was used by everybody and its contents were usually downed in one gulp. The Guinness often had flakes of chaff floating on the top but that didn't seem to bother anybody. I remember putting my nose to the top of the big earthenware jars in which the Guinness came and loving the smell of the hops and yeast. Dick Shanahan gave me a sip from his mug, but I didn't like the taste. Dick never helped out with the work. Instead he would sit nearby overseeing the whole procedure and we sat with him listening to his jokes and stories. He talked about Uncle Johnny's farm, to which there was no heir and tried to convince me that I would inherit it because I was called John. I think I wanted to believe this. The status of land ownership was highly regarded in rural Ireland of the nineteen-fifties and even as young children we were subconsciously aware of it. Snobbery was alive and well amongst the farming community and there was no avoiding its contamination. When the work was done Mammy had a huge pile of boiled spuds on the table and the whole *meitheal*, ten or twelve hungry men, sat down to a dinner of bacon and cabbage and mugs of milk.

But the days of the thrasher were numbered. Times were changing fast. A new invention called the Combine Harvester would end forever that exciting event that bonded a community together in our yard. Dick Shanahan's awe-inspiring machine would be consigned to history and left to slowly rot at the back of his farmyard. The Combine could do the cutting and threshing at great speed in one operation, manned by just two men, one driving and one tying the sacks of grain. Dan Mangan, who was a close friend of the family, had bought a Combine and Daddy stayed loyal to him, even though he 'drove us up the walls' on many occasions. He would make a start on a field of corn late in the evening and leave his Combine there on the understanding that the following day would see the job finished. But Dan was a martyr for the Guinness. On his way back from Lixnaw creamery he would start drinking and would sit there for hours on end, drinking pint after pint. With the corn ready and the sun shining, Daddy would be in a state of high dudgeon waiting for him to come. He would send Kevin on the bicycle to look for him. Kevin had great patience and would cajole and coax him until he agreed to come. The job would get finished and Daddy would never utter a word of annoyance. But one year most of the barley was lost because of Dan's drinking. This was more than friendship and loyalty could stand. From then on it would be the Roche brothers, Harold and Finbar, from Abbeydorney, who would cut our corn. They were two wild, devil-may-care young men but they were extremely good at doing the job and cutting the corn became a fun event once again.

Front view of Crotta Great House (circa 1902). Courtesy of the RIC archives.

Rear view of Crotta Great House (artist's impression).

Above: The demolition of Crotta Great House (early 1920's). Courtesy of the National Library of Ireland.

Right: The remains of the back entrance of the Great House still standing (1965).

Below: The remains of Crotta's cider industry - This massive cider press bears the inscription:
'R. Ponsonby 1760'.

Top: Jeremiah with his old hay-turner.

Left: Wedding photo of Jeremiah and Madge.

Above: Our family home (late 1950's).

Clockwise from top left:
Mick in his Sunday clothes with Dan and Diarmuid.

Mick white-washing the back kitchen.

Myself and Kevin in our home-made playsuits.

Minding Tom (in the pram) on a sunny day.

Top: Mammy with (l-r) John, Dan, Tom, Diarmuid and Kevin (1956).

Left: John and Tom (back), Margaret and Kevin (front).

Below: Mammy, Mick, Pat and Diarmuid in a field of corn.

Jeremiah and his six sons (with the ivy-clad orchard wall in the background).

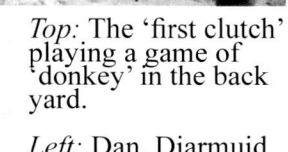

Top: The 'first clutch' playing a game of 'donkey' in the back yard.

Left: Dan, Diarmuid and Pat beside the Morris Minor.

Below: Dan Mangan's combine harvester.

Clockwise from top left:
Author's First Communion picture.

Uncle Danny's ordination photo.

Parish church, Lixnaw.

The old family plot and the new plot at Kiltwomey cemetery, Lixnaw.

Chapter Seven

The Old and the New

In the early fifties living conditions in our house were primitive. Household chores were difficult and laborious for Mammy and Sara. Cooking and baking were done on the open fire and clothes were washed by using the scrubbing board. The ironing of clothes with the heavy metal iron was slow and tedious. In winter, light or rather the lack of it was a problem. In the early days we had a large oil lamp suspended from the ceiling, casting scary shadows in all directions. It was a beautiful brass lamp, probably a relic of bygone days, but it only gave a dim gloomy light into the dark recesses of our big kitchen. A long chord ran from the lamp to a balancing weight on the side wall. This allowed the lamp to be lowered to add oil and to clean the globe before being re-lit and raised to its normal position. But semi-darkness was the dominant feature of our kitchen at night. A big blazing fire in the wintertime added a little extra light and candles were used for reading or sewing while a portable oil lamp was used for upstairs.

Then came the Tilley lamp. The Tilley lamp was an ingenious invention and we marvelled at its bright glow. Getting it started was part of the excitement. The container holding the oil had to be pressurised with a long pumping action. As our hands got tired we would take turns at doing the pumping. When the pressure was complete the oil was vapourised and as it sprayed onto a white mantle it gave a bright white glow that reached into all corners of the kitchen. Now candles were no longer needed for reading. But upstairs was still a scary place. By day, things were fine but during the night, if I chanced to wake up, I would keep my head under the quilt until I went back to sleep or else wake up Kevin to keep me company. There was safety in numbers. Kevin, Tom and myself shared a double bed, with Tom filling the middle slot. The mattress was a bit hollow in the centre and the pull of gravity dragged us in that direction. Sometimes Tom woke up in a fighting mood with both elbows flailing, complaining that we were sandwiching him.

For most of the year the house was cold. Even though the fire was never let go out it failed to make the house comfortable. The first job Mammy did each morning was to light the fire. Usually there were enough hot cinders in the hearth from the previous fire to get it going quickly. She would pile on the sods of turf and then fill the big black kettle. The kettle, hanging on the crane over the fire, took a long time to boil. Breakfast would have to wait and anyway there were many other jobs to be done first. After breakfast she would lift out some hot cinders from the fire using the long iron tongs and place them in a circle under the bread pan. When the pan was hot she would place the dough inside. The dough was mostly made of white flour with a small amount of brown flour added for roughage. It was circular in shape, about two feet in diameter. When she had put the lid on the pan she would place many hot coals on top to ensure the bread got fully baked. On busy days she baked two such loaves and most of it would be eaten by nightfall. It was the custom in our house to have tea and bread at eleven o'clock in the morning and at four o'clock in the afternoon. After supper at night, while we sat around the fire, Mammy would still be working. Her jobs were never-ending and we were blissfully oblivious of her workload.

So Mammy wanted changes. Daddy was dubious. Money was scarce and hard earned but Mammy had a glib answer that seemed to work: "*One more sow, one more cow and one more acre under the plough*". And so it began. In the summer of '56 our lives were made more exciting by renovation work on the house. John Herbert, a carpenter and builder of a daring disposition, took on the difficult job of upgrading and modernising what had been put in place three hundred years earlier. Mammy became the driving force; she wanted a range instead of the open fire. Some weeks previously she had gone to Tralee and put a deposit on a Stanley No. 9, which had a lovely turquoise ceramic finish, with a drop-door front to let the heat out.

To accommodate the range the enormous cavern of the old open fire place had to be made small by building a concrete wall on either side and at the back. When it was smoothly plastered the whole area was finished off with a cream coloured tile making it look very modern. Daddy's long box, that cosy seat by the open fire and a treasure trove of all sorts of bit and pieces, was consigned to the scrapheap and was replaced by two slender presses on either side of the newly fitted range. Jack Crowley, the local blacksmith, made an iron rack that was fitted over the range for drying clothes. Mammy painted it with silver paint and was pleased with how the whole thing looked. The range was clean and tidy and more importantly it made cooking and baking much easier. It radiated great heat when its large firebox was packed with black turf and for those who got the prime spots in the family circle, where long legs could reach the flat ledge along its front, it was luxury beyond belief on cold winter nights.

When the kitchen was finished John Herbert took on a more daring task. With just hammer, chisel and crowbar he broke through the four-foot thick wall in the downstairs bedroom, making an opening big enough to fit a six by four window. He then repeated this task in the room directly above. Steel framed windows were the 'latest thing' so these were fitted and the gaping sore of the wall plastered over. It made a great difference to the two rooms, letting in the morning sun. There were no east facing windows in these rooms because the protruding wing of the Great House had overlapped at this point and when that wing was demolished it left the remaining wall with huge rocks and stones jutting out. John Herbert proceeded to build up the outside of this jagged wall with solid concrete using rough planks of wood as a casement. The work was slow. Daddy and Mick mixed the concrete with shovels and climbed up the ladder to pour the bucketfuls into the casement. It was difficult work but week by week the casement was rising higher.

On the last day when he was almost at the top, disaster struck. The props they were using were not long enough for a good angle of support. The weight of the concrete and the law of gravity combined to bring a whole day's work crashing to the ground. But John Herbert would not be beaten. He extended the length of the props by nailing several long planks together and refitted the casement. But try as he might he couldn't get it back flush with the existing wall and when he refilled it from the heap of fallen mortar on the ground it left a big bulge. However, the job got finished without further mishaps and when the whole wall was finally painted it looked fine.

John Herbert had a great sense of humour and he always engaged with us as children. We liked hanging around watching him at work and his witty comments made us laugh. On the last day of school we ran all the way home, delighted by the thought of a whole summer of freedom and eager to see what work he had done. He asked us how long we got for the summer holidays and we told him "Seven weeks and three days". His response left us a bit puzzled. He said, with a deadpan face, "When you get over the three days, you won't feel the seven weeks".

His final job was very dangerous. For some reason the two chimneys in our house were built exceptionally tall, about seven feet. They were built of red clay brick and had stood proudly erect for three hundred years. But the previous winter a strong storm had blown off a few bricks from one chimney, smashing some of the slates as they fell. So this intrepid builder climbed onto the roof and walked along the top to survey the damage. It was scary looking at him standing so precariously on the ridge tile without anything to hold. The state of the chimneys was worse than he expected. The old lime mortar was crumbling everywhere and all the bricks were loose. He explained the situation to Daddy and said the best thing to do was to reduce the chimneys down to a safe level. Daddy was worried; demolition work on this house had already claimed the life of one man, but John Herbert dismissed his anxieties with a laugh. The following day, armed with a lump hammer and a chisel he climbed up, roped himself to the chimney stack and brick by brick removed about five feet from the first chimney and then repeated the same operation on the other one. Later, with a bucketful of cement mortar he climbed back up and secured the remaining bricks and made the chimneys safe. Our house had lost some of its historic grandeur; the proportions seemed wrong and the new chimney stumps looked strangely out of keeping with the rest of the building.

Mammy was on a roll and she wanted more. Next came the electricity. The rural electrification scheme had been rolled out all through the fifties and at first the power line had passed us by without a connection, but eventually Mammy had her way and the money was found. We watched with excitement as a gang of men erected five tall creosoted poles across our fields and then John Joe Quilter was contracted to wire the house. He put a single light-fitting in each room and a two-pinned socket in the kitchen for the new copper-coloured electric kettle. The single socket was half way up the wall over the kitchen table and from this he ran a spur to the Sacred Heart lamp. This little red glowing cross had become an essential part of the electric wiring of every good Catholic household throughout the country. Above the little red cross hung the picture of Jesus with His exposed glowing heart, which now glowed brighter. There was no more need for priming the Tilley lamp; with the throw of a switch the hundred watt bulb in the middle of the ceiling eradicated all shadows from even the darkest corners. Ghost stories would no longer be quite so dramatic and scary and the Christmas candle to welcome the Holy Family would be consigned to history; it was easier and safer to switch on the light in all the rooms.

But Mammy wasn't done yet; the following year brought the 'running water'. Ever since we were strong enough to lift a bucket, our first task after school was to go to the well for two buckets of spring water, a distance of about five hundred yards. The full buckets were very heavy so Kevin being the strongest would go in the middle holding the bucket handles in a certain way. Tom and myself would grab the outer side of the handles, leaning outwards as we walked up the road. Sometimes we jerked the buckets filling Kevin's shoes with water and he'd be hopping mad all the way home. But all of that was to come to an end; P.J. Stack took on the job of 'laying in' the water. Daddy and Mick built a small concrete shed with a tin roof which was called the pump house. A two-inch plastic pipe ran from the well to the pump house where the water was pressurised. From there it served five taps, two indoors for household use and three outdoors for farm needs. Mammy now had a large white ceramic sink in the back-kitchen for washing dishes or for washing clothes. But within a few months she had installed a Speed Queen washing machine, which, as well as washing, had a hand-operated ringer for squeezing out the clothes. Out with the old and in with the new! Washing clothes was now less laborious but was still tedious and time consuming. Sometimes we would help her feed the wet clothes through the ringer. The ringer was tight but sponge-padded and we would test our pain endurance by putting in our fingers with the clothes.

This was the end of Mammy's modernisation plans for the time being. Her plans for a bathroom and toilet would have to wait for another ten years. The great outdoors served as a toilet and large dock leaves were in ample supply. A large circular boiler served as a bath for us as children every Saturday night. Daddy had fitted this into our back-kitchen many years earlier for boiling spuds and vegetables for the pigs. It consisted of a heavy steel tub that sat in a cast iron surround, under which was a large firebox. So on Saturday nights this would be filled with water from the newly installed tap; a big fire set underneath and then allowed to cool off until it was safe to get into; rustic but effective. It was like a modern day Jacuzzi with three excited children splashing each other and making waves. One by one we'd be called in to the kitchen for hair washing over a basin placed on a chair. Mammy always used the green Loxene shampoo which was deadly on the eyes when you were being rinsed off. While all that cleaning and scrubbing was going on, Sara sat by the shoe press diligently polishing our Sunday shoes for Mass the following morning.

Life had become a little easier for Mammy and Sara. Household tasks were less arduous and more convenient now. Baking was certainly easier. The new Stanley range oven could fit two baking pans. When baked the loaves were placed on the small table by the front window, filling the kitchen with the delicious aroma of freshly-baked bread. Preparing dinner was also easier. Our dinner was always at one o'clock because Daddy liked to listen to a radio programme called the *Kennedys of Castleross*. Sara and Lizzie would have their dinner at the small table by the front window and have their own conversation. On Fridays Sara would read *The Kerryman* with the help of a magnifying glass while picking bits of crust from the fresh loaves on the table. The rest of us would eat our dinner at the big table at the other side of the kitchen. A big dish of boiled potatoes got centre stage and our plates would have generous helpings of bacon and cabbage. I was a poor eater. I liked the cabbage when it was a mixture of white and green cabbage and I would only eat the lean bacon. Our drink was always the same: a mug of milk. One thing the dinner table always provided was entertainment, especially from the 'younger clutch', Dan, Diarmuid and Pat. Dan would not drink the milk unless he got it in his "own puppy mug" and always kicked up a fuss if somebody else was using his mug. Diarmuid wouldn't eat his mashed potato unless he was given his "own big grey fork" and threw a tantrum if it couldn't be found while Pat struggled with his spoon getting more food on his face than in his mouth.

One day a neighbour rambled in at dinnertime and was invited to sit down and join us. He lived alone and was a bit innocent in his ways. What he said was never taken seriously. Mammy knew he had come expecting to be fed and gave him a big plateful of bacon and cabbage. He was sitting directly opposite her and was wolfing down his dinner with great gusto. Tiny bits of cabbage hung to his unshaven chin and he talked with his mouth full. She, on the other hand was eating very slowly and very carefully, picking the softest bits of food with her fork. During the course of conversation he asked what was wrong and she explained that her false teeth were hurting her gums. Without hesitation he took his own false teeth out of his mouth and said "try mine, they might fit better". After a moment of shock the table erupted in laughter as Mammy declined his well-intentioned offer. He replaced his teeth and continued eating but she had lost her appetite and went to boil the kettle for the wash-up.

Our kitchen table held so many memories. If tables could talk what a story it could tell. All through the years it occupied a prime spot beside the back window. It was very solid and therefore had many uses other than a kitchen table. It was used as a bench on which pigs were killed outside our backdoor. It was used as an altar for the priest to say Mass, when our turn for the Stations came around. It was used as a workbench for various kinds of jobs. It was used as our desk for doing homework every evening after school and on Sunday afternoons Daddy would lie on it listening to a football match on the radio.

The radio was located nearby on the window sill. We would all gather round listening to Micheál O'Hehir describing the exploits of the great Kerry footballers or Tipperary hurlers. It filled our heads with youthful imaginings and when the match was over we would grab our hurleys and attempt to emulate the skill and valour so vividly described on the radio. But Mammy had no interest in Gaelic games and would sometimes go to bed on Sunday afternoons. Other times she would work away on her Singer sewing machine. She was a dab hand with this machine and when Kevin, Tom and I got the 'call up' for the Crotta juvenile hurling team, she designed and made up our hurling shorts from an old not-so-white bed sheet. This bed sheet had been made from large cotton flour-bags that she had sewn together. It had been bleached white but it still carried a few printed words in places. We looked a bit odd and were very self-conscious of our homemade shorts and only wore them one time.

Chapter Eight

Hurling

Hurling is Ireland's ancient game, supposedly dating back to the legend of *Cú Chulainn*. Perfecting *Cú Chulainn's* skill with the hurley was my boyhood ambition. He was my hero. Even as a young boy he could strike the ball high in the air, run forward at great speed and strike it again before it touched the ground. On one occasion when a fierce wolfhound was charging down on him he struck the ball with deadly force and unparalleled accuracy into the hound's open mouth killing it instantly. Hurling was the stuff our dreams were made of. As small children we even played hurling in the kitchen, but only on nights when Daddy was gone to the rambling house. He would go there a few nights each week over the winter. We knew how to take advantage of his absence. Some nights we'd start complaining about being cold and sure enough Mammy would say "Go, play hurling".

The two large doors on either side of the kitchen, well recessed into the thick walls, were perfect as goalmouths. The ball was usually a paper ball bound tight with a piece of twine, so it was relatively safe to strike it hard. The battle would rage across the concrete floor until we were exhausted or until we got a bit too boisterous for the adults who were sitting around the fire. Sometimes Sara would complain saying: "Go easy, ye're raising too much dust" but the game would continue unabated. On one occasion, in a desperate effort to score a goal, Tom took a wild swing at the ball and smashed the glass of the Sacred Heart picture that was hanging on the wall. Consternation and reprimand ensued and that put an end to hurling in the kitchen for evermore.

As we grew older our love of hurling grew stronger. On summer evenings when asked to round up the cows for milking I would always take my hurley. I would strike the ball high in the air as *Cú Chulainn* had done and try to strike it again before it touched the ground, but without much success as the cows always got in the way. On rainy days we sometimes played hurling in the front stall. The space was a little confined but at least we had no worries about breaking holy pictures. This made the games more vigorous and careless. On one such day I broke Kevin's front tooth, his adult tooth, the left incisor. I felt really bad about my careless stroke but he made no big deal of it. For weeks it played on my mind and I couldn't stop thinking about it. I knew I had left him marked for life and there was nothing that could be done.

But our passion for hurling continued. When we weren't helping out with farm work we would grab the hurleys. There was no question of special gear, or boots, or even proper hurleys. Often we used badly shaped homemade hurleys or badly repaired ones that were lethal in a clash, but most of the time we escaped with only minor scratches and skinned knuckles. We played in our special pitch in a corner of the back-lawn field. We had erected two sets of goal-posts from straight ash branches and tied a length of binder twine across the top as the cross-bar. This helped in the hotly contested arguments of whether a particular shot was a goal or a point or wide. But these games were only a warm-up for the evening matches in Stack's field.

Young and old for miles around would descend on Stack's field every evening. The age range went from Johnny Kenny at five to Mick Falvey at seventy five. Starting times and finishing times were loose. The match might be in full swing but late arrivals joined in at will, pairing off indiscriminately, often resulting in very bloated and uneven teams. It was good fun but serious stuff. Winning was important and showing off one's skill even more so. This was self-regulated hurling without the services of a referee. Very often 'might was right' and 'dirty tricks' were par for the course, but one evening Tom could take it no more and threw down the gauntlet to Danny Lawlor. Danny was the sort of player that nobody wanted to mark. He was rough and awkward and ruthless and had a favourite expression, 'hit him on the backbone' for discouraging young players. Tom threw down his hurley and challenged him to fight, much to the enjoyment of the older players who watched in amusement. One of them shouted out: "Go on, Johnny Galvin, stand up to him". Many years before, our uncle Johnny had been a star player with the 'Tullig Gamecocks' hurling club and had a reputation for being a bit feisty. So Tom was the hero of the hour. It was better than scoring a winning goal. But there was no fight, Danny was a good sport and apologised and so the game continued. On those warm summer evenings the match would go on till we could no longer see the ball clearly and then we would drag our weary limbs towards Moira's shop, dehydrated and gasping for a drink. Most of the time it was just water but sometimes we'd have the price of a bottle of lemonade but this had to be shared with other thirsty mouths. The orange ice-pops for a penny were also in huge demand. They dissolved slowly in the mouth and tasted delicious.

A fist fight broke out one evening between two of the older players just as the teams had been picked and the match about to start. One had been in the army and had developed some boxing skills but the other was taller with a longer reach. We stood perplexed as we watched the duel unfold. The army man kept attacking aggressively while the taller man tried to avoid his punches as he backed away. Our spellbound gaze followed the fight as it circled around the playing field. Arms were flailing left and right as the army man kept up his aggressive attack. The tall man's nose was bleeding profusely but he occasionally connected with a fierce punch. The fight raged on and on encircling us as we stood there, stunned and entranced, by what was happening. Nobody intervened. The atmosphere was becoming strangely unreal and we had all gone silent as an older man entered the field. He was Tom Nolan, six foot two and built to match. Down through the years he had been Crotta's most outstanding player, one everybody admired and respected. He marched directly to the fighting duo and broke up the fight. He had to push them apart again and again, shouting "Stop, enough". The army man lashed out at him but Tom Nolan caught him by the throat and held him at arm's length till he calmed down. After a long pause the atmosphere returned to normal. The ball was thrown in and the game began.

The hurling match in the evening was the highlight of my day. No matter how tired I felt there was always enough energy to be found. One balmy summer's evening as we made our way down to the hurling field, I was in an excited mood. I had a new hurley that Patty Kenny had made for me. It felt good in my hand and I was dying to try it out. I kept visualising myself lifting and striking the ball with great skill. We stopped at Jack Crowley's forge as we were passing by. The forge had been the gate lodge during the Ponsonby era, near to where heavy iron gates had guarded the approach road to Crotta House. It was situated on the bank of a fast flowing stream and was also used as a laundry where clothes were washed for the landlord's family. It was now the local forge and the stream was used for cooling the big iron bands of cart wheels. It was always a place of interest and we were fascinated by the work of the blacksmith. He was a hard worker; being a postman by day and a blacksmith in the evening and was a good friend of the Galvin family. He was shoeing a horse as we passed by and this was something we liked to watch.

Chapter Eight

As he is shaping the piece of red-hot iron on the anvil sparks are flying in all directions. With a few deft taps of his heavy hammer the first shoe is ready. He asks me to stand by the mare's head and hold the reins so that she won't move. I'm delighted to be chosen and given a job of such importance. He presses the hot iron shoe against her hoof and this gives off a cloud of smoke and a pungent smell of burning. It's exciting to be there. Tom and Kevin are watching with glued attention as the first shoe is nailed into place. I'm rubbing the mare's soft nose and talking softly to her saying "It won't be long now". I'm so absorbed in what I'm doing that I haven't noticed them drift away to the hurling field without me. Suddenly I feel very much alone, abandoned and slightly afraid in the semi-darkness of the forge. Jack continues pounding fiercely on the hot iron and pretends to be unaware of my anxiety. I ask in a timid voice if I can leave and he gruffly answers: "No, you must stay until all the shoes are fitted". He's standing between me and the door and so I'm forced to remain there for what seems like hours. I ask again "Can I go now"? but he shouts "stay where you are" over the clanging sound of the hammer. I'm angry but I'm too frightened to do anything about it. I don't understand what's happening. He doesn't really need my help, but I'm too timid to protest. He's Daddy's good friend and I can't understand why he's doing this. My agitation and distress are plainly visible but he pretends not to notice. The time drags slowly. He seems to slow down his pace as I wait for permission to leave. Finally he says: "You can go now". By the time I get to the hurling-field I am too upset to join in. My evening has been ruined, so I sit on the sideline with Billy Kenny who is nursing a nasty blood injury he got during the game. Billy is my best friend and talking to him eases my mind. Together we watch the frenzied action on the field and discuss the skills of the outstanding players, as the game rages back and forth.

In 1959 Crotta hurling club won the juvenile county championship. This victory was an outstanding achievement in that Crotta was then a separate club from Kilflynn and had a very limited pool of players. Some families, ours included, had two or three team members. There were three Leens, three Galvins, two Kennys and two Nolans. One of the Nolans, Todd, was a master of the craft and could single-handedly turn a game around. He was a joy to watch and an inspiration to the rest of us. A word of praise from Todd was enough to boost my confidence for the whole game. It was very exciting to be caught up in the enthusiasm of getting to the final but also very nerve-wracking. We had won our four previous matches. At the start of a game I'd be jittery with nerves but if the first strike of a ball went well, confidence would take over and I'd play like a demon. But the opposite was often the case; fumbling that first catch of the ball would leave me fumbling every ball that came my way; missing golden opportunities and hearing a lot of derisory comments from the sideline.

We were driven to various venues during the course of the campaign by the same five drivers who were staunch supporters of the club. We would pack ourselves into the back seat with reckless abandon. Overcrowding never seemed to bother our drivers. The same reckless regard was paid to safety on the field; the hurling helmet hadn't yet been invented and games were fast and furious. Over-enthusiastic supporters would shout all sorts of dangerous incitements at young impressionable players. In the midst of flailing hurleys I would sometimes hear a supporter shouting, "Galvin get into it", while I was doing my best to stay out of it. Another voice could be heard telling a player, whose opposite number was standing on the ball, to "Cut the legs from under him". Such lunatic encouragement could be heard at all our games. The over-riding need to be victorious blinded supporters to the dangers and possible disastrous consequences for the young innocent players involved. Luckily most of us escaped with just cuts and bruises and the words of praise ringing in our ears made the risk worthwhile. Sweaty and dirty we got back into our clothes at the side of the field and piled into the cars to be taken for chips and lemonade before setting out for the journey home. We always travelled in Michael Kenny's Ford Anglia because Mammy knew he didn't drink and believed he was a safe driver.

No other match generated as much rivalry as a game against Lixnaw. This might have been due to the animosity that sometimes arises between neighbouring clubs or maybe it had its origins in the nineteenth century faction fighting promoted and encouraged by the Crotta landlord, Thomas Ponsonby. Whatever its origin it was deep-rooted. On one occasion during a senior championship game the rivalry boiled over and the game became a free-for-all. I watched in disbelief as the melee broke out. Suddenly, two of the outstanding hurlers on the field, Eamonn Leen and Richie McElligott caught everybody's attention. They were swinging their hurleys at each other with brute force and ferocity. Both were big men and evenly matched and not an inch was given by either; it was like a scene from a Viking battle, with the hurleys as broadswords. The fight raged backwards and forwards, both had drawn blood but neither would give in. All other activity on the field had ceased as the crowd gazed, in dazed silence, at this 'death duel'. Eventually, after what seemed like an age, some brave volunteers ventured in to break it up. The moment of madness had passed off without any serious casualties but feelings were still running high. Not surprisingly, the referee called off the match and the players were kept apart when leaving the field; there had been enough faction fighting for one day.

We never played hurling at school. There was no enthusiasm for sport amongst the teachers and no sport equipment of any kind was ever provided. There was no playing pitch available, but in the autumn of my fifth class year John Weir gave us the use of a field beside the school. It was totally unsuitable for hurling as it was a tilled field from which he had harvested a crop of barley. The stubble was much too high for ground hurling and the stony surface presented many challenges but our enthusiasm overcame these difficulties. So every evening at three o'clock we would hop over the school wall and get a game started. Two sixth class boys would pick the teams, taking it in turn to call a name. Each of us waited in anticipation to hear our names being called and as we listened our status in the pecking order of hurling was becoming clear. I was never first pick, always around fourth or fifth, but would be happy to find myself on the same side as Billy Kenny or Donal Leen.

I had a new home-made hurley that felt very light in my hand and I could solo with one hand like a senior player. It had been roughly shaped by Uncle Pat from a piece of dried sycamore but it didn't have much of a curve at the end. It looked more like a wooden spade than a hurley but this made it easier to lift and strike the *sliotar* (hurling ball) in the stubble field. I could strike left or right with ease. My skill was developing and my confidence was growing. It was good to be part of this crazy melee in the stony stubble field after school. We played with a lovely light *sliotar* that John Shanahan had purchased in Tralee. We each had given him 'trupence' the previous week to make up the price. I liked the feel and the sound of my hurley striking the ball. I carried the feel and the sound in my head as I walked home after the game, re-living every exhilarating moment. We had been blessed with many weeks of dry sunny weather and our stubble playing pitch had become smoother and more solid from the trampling feet. It was now more suitable for ground hurling, which made our games fast, furious and fiercely exciting. John Weir's ploughed field had made school a more attractive place. Each afternoon, I waited, impatiently, for the school clock to strike three.

Chapter Nine

Christmas

Older people would often say to us "Your schooldays are the happiest days of your life". I didn't want to believe it because it held out such a gloomy prospect for my future. But the holidays almost made up for the bad times, especially the Christmas holidays. The thought of Christmas made first-term bearable. I could live on the excitement that lay ahead and it was always worth waiting for. With mounting anticipation, I counted down the days. In the build-up to Christmas everybody in our house played their part. There were many cypress trees growing around the house and Daddy would cut a few branches and tie them together in the shape of a Christmas tree. This would be secured in a butter-box filled with sand and placed at the back of the kitchen. Mammy would cover the box with red crepe paper and would make up a flour paste which she would smear onto the green branches to make the ends look white. The crib would be given pride of place on a small table beside the tree. The colourful chalk figurines in the crib took up their usual places where they would remain until the day after Little Christmas.

From years of following the same routine we knew exactly where every Christmas decorations should go, each thumb-tacked in the correct place on the walls or ceiling. If the holly produced its berries early in December, Mick would cut a few branches and store them in a dark place in the hayshed. He would retrieve them the day before Christmas Eve and we would place small red-berried branches on the window sashes and behind the pictures on the walls, giving our kitchen a fresh green look. At the start of Christmas week Mick would sweep the backyard from end to end with the strong coarse brush. This was a difficult job because most of the yard was covered in several inches of slush and dirt and the surface was very uneven. When the sweeping was finished he would begin the whitewashing. Armed with his big bucket of lime wash and a white-washing brush he would whiten every patch of wall about the yard that needed a touch up.

Early on Christmas Eve, Mammy, Sara and Lizzie would take us to

Lixnaw church for confession. Daddy and Mick weren't quite so devout and only went to confession once a year; that was at Easter time to do their Easter duty. It was a Church commandment that all Catholics do their Easter duty by going to Confession and Communion at least once between Ash Wednesday and Trinity Sunday. So Daddy remained behind to baby sit and was given instructions about decorating the Christmas tree, but instead of taking down the box of decorations he hung kitchen utensils on all the branches. We came home to find the tree covered with pots, pans, brushes, towels, and his old cap on top where the angel should be. It was a rare moment; he had taken time out to do something silly and funny; he had let down his guard and allowed the 'inner child' a moment of freedom. We stood there wondering how Mammy would react but she just burst out laughing.

Some years our Christmas was made extra special by the arrival of a large pre-Christmas box filled with toys and books. In some respects this package of surprises stole the show from Santa's surprises. This parcel would arrive from London, sent by Lil Shanahan, and was filled with all sorts of interesting and exciting toys. We didn't know then why Lil Shanahan sent these parcels and didn't ask, we just gazed in wonder at the colourful contents. In later years we discovered why. The Shanahan family from Crotta Cross had all left home, some had gone to London and some had settled in Dublin. Their father was old and alone, but continued working the farm and Daddy helped out with the work and looked after him for many years. The parcel was his daughter's way of acknowledging this and showing her appreciation. She obviously didn't select and pack the toys herself; most likely placed an order with some toy store who posted it on, because on one occasion toy cricket sets were included with a real cricket ball. We were amused by the red wooden ball and couldn't understand how anyone could play with something so heavy and hard. We knew nothing of cricket and cared less. But the two tennis balls were a different matter, so white and furry. I claimed one of them as my part of the booty and kept it hidden in a secret box. I seldom played with it in case in might get dirty and only rarely took it outside.

The week before Christmas Mammy went shopping to all her usual places to make sure she received her 'Christmas box'. This took the form of a free tin of biscuits or sweets or some other item and was the shopkeepers' gesture of appreciation of her custom throughout the year. The final shopping trip was to Jack Mac's pub for a crate of bottled Guinness and maybe a half bottle of whiskey. These were not for family consumption but hospitality demanded that visitors should be offered something stronger than tea. Christmas Eve preparations filled the kitchen with tantalising aromas and the promise of good things to come. Dissolving the jelly in boiling water for the sponge trifle perfumed the house with the sweet smell of strawberries and raspberries. When Mammy wasn't looking we bit off pieces of the concentrated jelly to suck as a sweet. Who could resist it? While stirring the trifle mix we eyed the three dome-shaped Christmas puddings on the shelf of the dresser; their dark brown richness teasing our taste buds. On the other table the stuffing mix prepared for the goose or turkey was scenting the kitchen with the savoury smell of herbs and onions. But these delicacies were for tomorrow. For now we would have to make do with biscuits and raspberry cordial. We went to bed early on Christmas Eve to make sure we were sound asleep when Santy (that's what we called him) arrived and because early Mass on Christmas morning was at eight o'clock.

One Christmas brought a hard frost. The chill of the early morning air felt cold and reddened our faces as the pony and trap made its way to Lixnaw. Daddy slowed down the pony whenever there were icy patches on the road. Ice crunched under the wheels of the trap as our gallant little pony trotted along. As we neared the village other pony and traps were pulling in to Jack Mac's yard. The wall beside Jack Mac's pub had a number of iron rings for tethering horses and here our pony would remain till Mass was over. As we walked to the church door, shouts of *"Happy Christmas"* assailed our ears from every side, to which we replied: *"Many Happy Returns"*. Good tidings and good will were in plentiful supply on this joyous morning. The church, though cold, had taken on a relaxed and happy atmosphere. The high soprano voices of the nun's choir filled the air as we waited for the priest to come on the altar. They sang *Silent Night*, *Away in a Manger* and the *Adeste Fideles*. The sound of the Latin words had a melodious quality even though I had no idea what they meant. Then a bell rang and Tom Carroll, the server, came out of the sacristy followed by Fr. Scanlon. When the Mass was over Mammy took us to the side altar to see the crib. The crib was an impressive sight with statues almost as big as us and I gazed in awe at how real they looked. Baby Jesus was loosely clad in his swaddling clothes, his arms and legs were bare. He didn't seem to mind the cold, but my toes were painfully numb in my shoes.

A week before Christmas Sara had put long red candles in each window in all the rooms. They were lit each night, as a welcome sign, in remembrance of Mary and Joseph's unrewarding search for lodgings in Bethlehem. The candles stood in large jam jars filled with sand, with red crepe paper wrapped around each jar. Every night, it was our job as children to check on the candles at regular intervals. This meant going upstairs and walking along an eerie corridor, where the armour-clad ghost of Henry Ponsonby might be lurking in the dark. We took it in turn to do the rounds. This job used to petrify me but I couldn't admit that. On the false pretence of caring for my younger brother Dan, then a toddler, I would ask him to accompany me. Somehow his childish conversation helped to reduce the terror of opening each bedroom door and peering in. The rooms were dark hollow voids and who could tell what ghostly things might be lurking in there. As well as that my parent's room had a black and white picture of the crucified Christ, his face distorted in agony that sent shivers down my spine. In the second room there was the large statue of the Virgin Mary, adorned with luminous glowing beads, looking ghostly in the dark. I tried to hide my fears from Dan by whistling a tune as we hurried back to the safety of the kitchen to report that everything was fine.

When I was ten, Santy brought us a bicycle; one bicycle that we had to share between the three of us, the older clutch. The younger clutch got their own individual presents. Even though Tom was still a believer he was a bit dubious as to why Santy would bring a second-hand bicycle and worse, a girl's bicycle, to three boys. But, that aside, it was in good working order and we were thrilled with it. It marked a mini turning point in our sheltered existence in that we could now venture a little bit further from the nest. Kevin mastered the balance very quickly and patiently helped myself and Tom to develop the skill. He would take us for cycling lessons on a stretch of road beyond the front field that had a slight incline. With his hand gripping the saddle he would hold the bike steady while running along beside, offering encouragement and advice to the learner. I remember on one occasion shouting at him to let go the saddle, that I could do it by myself. When we got to the end of the road I said "Why didn't you let go" and then he told me that he hadn't been holding the saddle at all for the last two runs, but was ready to catch me if I wobbled. I couldn't believe it, I could cycle: a major achievement. I was so elated that I continued doing the same run until it got dark.

That was the last year Santy came to us, the older clutch. The following year hopes were still high, even though we were well beyond the age of belief. In our house Santy actually delivered the presents into the bedroom and placed them on the quilt at the end of the bed. On this Christmas morning we woke to the sound of rustling paper as we moved our legs. Wow! Fantastic! Santy had not forgotten us after all. With rising excitement we began to tear open the boxes, but Mammy had played a cruel trick on us; our three boxes were filled with pulped turnips. The disappointment was heartbreaking and even though she had new jumpers and socks for us when we got downstairs, the anti-climax hung over our whole day. I think she regretted her thoughtless prank because the following year she involved us in the secrecy of selecting the presents for the younger ones. On that Christmas Eve we stayed up late and helped to wrap up the gifts, while constantly keeping a listening ear for any footsteps in the room above. Then, with tingling excitement, we quietly tiptoed up the stairs and sneaked the colourful bundles onto their bed.

St. Stephen's Day was known as the 'Day of the Wran', when a little wren, the king of the birds, was killed, tied to a holly branch and was carried by the Wren Boys as they made their way from house to house. The wren was killed, not because he was king of the birds, but because he had betrayed St. Stephen when he was hiding from the Roman soldiers. The wren was chirping so loudly on a branch that it drew the attention of the soldiers to his hiding place. St. Stephen was then taken away and stoned to death. We used to sing a rhyme that went: *The wran, the wran the king of all birds, on St. Stephen's Day he was caught in the furze.*

The 'Day of the Wran' was a day of merrymaking and enjoyment with music, song and dance at each house, accompanied by the occasional drop of whiskey or Guinness. It was exciting to see an oddly-clad group of entertainers come into our kitchen with masks on their faces. Mammy's curiosity would get the better of her and would cajole them into disclosing their identity. Then the singing would come to an end and the tracing of relationships would begin. The 'wren boy' custom was popular amongst children and young adults and was seen as a means of getting some money. Daddy saw it as a form of begging and would not allow us to do it, but one year he relented. We decided to chance it despite our shyness. We made face masks and put on odd looking clothes and set off. Martin Crowley came with us to give us moral support. Bridie Gallagher had recorded her very popular song *The County of Armagh* and this became our 'party piece' at each house. We realised that having our faces covered was a big help when trying to sing in public. Painful as our rendition surely was, we were amazed at how good humoured and generous the people were. By the end of the day our little moneybag was heavy with silver coins which we divided equally between us.

But all too soon the Christmas holidays were over and we were back at school. Kevin, being the eldest, had first claim on the bike, so every school morning he would head off on his own without a thought for myself and

Tom. When he got near the school he would hide the bike at the back of Mick Shanahan's cottage and walk the remaining bit. Our little second-hand bike was far too precious to risk loss or damage. But it reverted back to joint ownership in the evenings. The joy of cycling was addictive. I would volunteer to go to the shop whenever Mammy was short of something for the supper. Back the Garrynagore road to Moll Grady's shop, flying at top speed especially whenever the slope was with me. There would always be a few pence in change at the shop, which I was allowed to spend on ice-cream. Moll had a generous heart and the tuppenny wafer I'd order would be much thicker than the tuppence merited.

On these excursions to the shop I would proceed further back the road to Donal Leen's house. Donal was a classmate and a close buddy who always made me laugh. He was everything I wished I could be. He was definite about things and was able to stand up for himself regardless of the consequences. The reason for cycling to his house was to swap cowboy comics. He always kept a good stash. I was totally hooked on Roy Rogers, Davy Crockett, Kit Carson, Buck Jones and of course the Lone Ranger and Tonto. As I cycled home the bike became my horse and I was the Lone Ranger shouting "Hi! Ho! Silver".

Chapter Ten

The Morris Minor

The bike had pushed out our boundaries a little bit but the purchase of the Morris Minor broadened our limited horizons and opened up our lives to many different experiences. Times were changing. Black cars, mostly Ford Prefects and Morris Minors, were becoming a common sight on the quiet roads of North Kerry. Sometime in 1955 our lovely pony and trap was replaced by a motor car, registration number ZX 832. Tom Rice, the main car dealer in Tralee, was a friend of the family and volunteered to give free driving lessons to Daddy as part of the deal. He would arrive a few evenings each week for the driving lessons and we enjoyed the unique thrill of getting a spin in our new car. The smell of the red upholstery on the seats and the feel of the soft velvety carpet on the floor made it all very exciting, but there was also a little bit of apprehension as Daddy struggled with the gear stick and clutch pedal. Even though the lessons went on for many months he never achieved full mastery of it. Clutch control and gear changes were a continuous problem.

But he didn't lack determination. He would take us on trips to Tralee or Listowel and Lixnaw on Sunday mornings. Reversing was difficult for him, so on the Sunday Mass trip he would do a u-turn in Jim Lynch's front yard and swing across the road and park it there. One Sunday afternoon he took the whole family to visit the Leens, my mother's side of the family, who lived by the sea at Meenogahane. The passageway into the Leen's house was very narrow with a right-angle turn off the road. He misjudged the turn and smashed into the corner of a wall. Shock and consternation descended upon our happy occasion. The day was ruined and so was the car. The left front wing and bonnet were badly damaged and the trauma and drama that followed left its mark on all of us. Ever afterwards whenever the car was turning in any narrow passageway I would put my head down and close my eyes until the car had come to a stop.

Next in line for a smashing was the headlight on the right front wing, but this time it was Kevin that did the damage. Daddy had constructed a rough-and-ready shed in our lower yard against a wall that in times past was a warehouse for the cider-making industry. This became known as the 'car house'. The corrugated roof was supported on heavy oak uprights with just enough room for the car to fit through. Most of the time the car was never locked and so was a big attraction for us to while away our time, sitting in it. Kevin loved to sit at the steering wheel pretending he was driving. He would forcibly push the gear stick through all the gears and keep turning the wheel to left and right. He was obsessed with steering, so one day we decided to push the car out of the shed, up along a slight incline in the yard, so that he could steer it back in to its house again. He sat in and Tom and I started to push. With the help of the slope it gathered momentum quickly and Kevin's steering wasn't up to scratch and he crashed it into the wooden pole, smashing the headlamp. We gazed in horror at the damage but when the shock had subsided Kevin decided that confession was the only option for us, so we came up to the kitchen to 'face the music'. Cleverly, Kevin pushed Tom ahead of him in to the kitchen to do the talking while we followed behind. Tom must have had an acute sense of how reverse psychology works because his opening sentence saved the day: "Mammy, we know you're going to kill us and we know we deserve it 'cause we've done a terrible thing". She must have expected something really bad because her response surprised us: "Don't worry about it, it can be fixed".

Daddy persevered with the driving and even though he had improved he still lacked confidence in himself. But he wasn't for quitting and planned a trip to the Listowel Races. This was a three-day racing festival in September and all schools in North Kerry had permission to close for the middle day. Our excitement grew as the day approached. Mammy, who had to stay at home babysitting, waved goodbye as Tom and myself clambered into the back seat and Kevin into the front. We got there early in the afternoon where Daddy met with cousins he hadn't seen for many years. Their names were Molly, Liz, and Lil and, looking at us, they greeted him with "You'll soon have help" and then Molly reached in her bag for a half-crown for each of us. We knew the value of the half-crown; that beautiful heavy coin with the horse on one side. But there was a whole ritual around the receiving of this gift. Daddy tried to prevent her from giving it, while telling us not to take the money, but eventually she pushed it into our supposedly unwilling hands. Jokingly, she admonished: "Don't put it all on the one horse" as we walked in to the race track enclosure. Even though we didn't win any bets, we had a long exciting day watching the horses clear the fences, but on the way home disaster struck again. Something happened during a gear change that jammed the gearbox. The gear stick was stuck and nothing could shift it.

Frightened and confused we push the car onto the grass margin of a lonely stretch of road and Daddy tells us to sit in and wait while he walks several miles back to Listowel to find a mechanic. We wait nervously in the growing darkness for what seems like ages. There's an eerie silence about the place and a pale moon is casting ghost-like shadows on the windows of the car. Suddenly, we stop talking. We can make out something coming towards us; it looms up out of the darkness, a strange and menacing shape. Kevin tells us to lock all the doors. As the clip-clop sound draws near we can make out the shape of horse and rider and a big dog running along behind. The rider slows down and looks us over and then moves on. We are growing more and more tense and secretly praying that Daddy will come quickly. Very few cars are passing by but as each approaches we hope it is him returning, only to be disappointed. We wait in silence, each of us lost in our own thoughts. Finding a mechanic willing to do a 'call-out' on the night of the races is a daunting task and must have been a disheartening experience but he succeeds. A car pulls up behind us and Daddy and the driver get out. The mechanic, holding a bright flashlight, pulls back the carpet and does something at the base of the gearstick to free it. We drive home silently, tension returning with each gear change. The gears grate and the engine chugs at every junction. It's a harrowing journey and I'm glad when the car comes to a stop in our yard.

Later that year when a memorial was erected in Knockanure to the memory of the men who had been shot by the 'Black and Tans', he decided he would take us to the unveiling ceremony. The well-known song, *The Valley of Knockanure* tells the story of four young freedom fighters, Walsh, Lyons, Dalton and O'Dea, who were hiding out in this area when they were spotted by a British patrol truck. Three of them were shot dead but O'Dea escaped with only flesh wounds and later went to America. We arrived in time to see the parish priest bless the small memorial stone in the field where they fell. He then gave a long moving oration. A small group of older men, bare-headed and silent, stood at one side with their heads bowed low. A local man with tears in his eyes said that they were classmates in their primary school days. Daddy talked with him for a long time about things that had happened in those troubled times. I listened with rapt attention to the stories he was telling about the daring exploits these dead heroes had engaged in and how the Tans had retaliated. The Tans were loathed for their callousness and inhumanity in their treatment of civilians and their destruction of civilian property. Daddy also had stories to relate; how Mick had been taken in to the barracks in Tralee for questioning and how Sara had been a member of *Cumann na mBan*; an

Irish republican organisation that had aligned itself to the Irish Volunteers, and how the Galvin homestead at Ballyrehan was sometimes used as a 'safe house' by some of the freedom fighters, local men who were members of the North Kerry Flying Column.

When the ceremony was over Daddy decided to visit his first cousin, Molly Shea, with whom we had spent the day at the races the previous year. Her home was not far from where we were and Daddy hadn't seen her since the funeral. Molly's husband, Michael Kissane, had died suddenly some months previously. She was still in mourning; all in black, and when they met tears were streaming down her cheeks. The laughing face that had advised us 'not to bet all our money on the one horse' was now clouded in sorrow. Her conversation was a series of stilted remarks in between the bouts of sniffling. Her four fatherless children, two girls and two boys, sat silently at the back of the room and we sat in silence with them not knowing what to say. In the midst of her grief Molly managed a wan smile and suggested we should go out and play football. The ball was old and slightly deflated but we picked teams and had an exciting game. The two boys, Richard and Frank, were the captains of our two teams and they lost themselves completely in the game, oblivious for a little while, of the sadness that surrounded them.

Another monument and another memorial service found us at Ballyseedy, near Tralee. Daddy's driving skills were showing a slight improvement but reversing was still a problem. His constant use of the horn and indicators amused us as we sat in the back seat. The indicator was a little yellow coloured arm about eight inches long that would pop out from the door frame when he pressed a lever under the steering wheel. He parked the car a long way off and we walked to where the ceremony was taking place. An impressive bronze sculpture had been put in place at Ballyseedy Cross, made up of three bronze figures - two figures depicting human suffering and the third depicting mankind's unyielding spirit. This spot had witnessed one of the worst atrocities of the civil war, something Daddy was quite emotional about because it involved three local men from Kilflynn: Stephen Fuller, George O'Shea and Tim Twomey. County Kerry had suffered a disproportionate number of vindictive killings during the troubles and hatred and bitterness were still simmering beneath the surface. After the Treaty was signed Irish public opinion split into two camps, for or against; the Free State side versus the Republican side. Friends who had fought side by side during the guerrilla years were now aiming their rifles at each other. In reprisal for the death of five Free State soldiers at Knocknagoshel, nine anti-treaty IRA prisoners were taken from

Ballymullen Barracks in Tralee to Ballyseedy Cross. They were tied together around a landmine which was then detonated, killing eight of them, including Tim Twomey and George O'Shea, but Stephen Fuller survived. He was thrown clear by the force of the blast over the nearby ditch. Though badly injured he crawled to a neighbouring farmhouse where he was kept hidden and nursed back to health. He had lived to disclose a heinous war crime; a calculated, cold-blooded atrocity carried out by the forces of the State. In later years when the republican party gained power he was elected to government and served as a Fianna Fail TD for north Kerry. Daddy and Mammy talked about these things as we drove home. They were both staunch Fianna Fáil supporters and their political views were closely aligned but both held strong views about DeValera's role in Ireland's struggle for independence and the subsequent civil war.

****

Our uncle Pat was home on holiday from London the following summer. He was Mammy's youngest brother and they were very close. For us he was an exciting big brother whose company we enjoyed because he didn't take life too seriously. He spent a lot of his time at Crotta and began learning to drive the Morris Minor; no worries about insurance or a licence. A workmate in London had asked him to bring home a parcel to his father who lived near Killarney, so Pat and I set off early one morning to deliver this cardboard box, tightly bound with strong twine, to its destination. I had never been further south than Tralee so this was a major adventure. When he got to Killarney, Pat didn't know which way to go so he had to ask for directions many times. Most of the people he talked to had never heard of the place we were looking for. He drove for miles along the Fossa road with the breath-taking beauty of the McGillycuddy Reeks in the distance.

After many more enquiries and many narrow bye-roads we came to a cottage in the shadow of a towering purple mountain. A glistening waterfall was splashing into a small lake on one side of the cottage and two small fields on the other side held a number of sheep and a pony. Two dogs started to bark at the sound of the car and an old man appeared at the door, studying us with a wary gaze. Pat called out to him from the distance and explained the reason for being there. He took the box out of the boot and we walked to the house. They talked for a while at the door; Pat telling him some story about his son in London and then he invited us in to his very small and primitive dwelling. The house had no electricity and very little in the way of comfort. He was living alone in abject poverty with only the dogs for company. Pat remarked on how beautiful the scenery was and he replied "What good is that to me". He was obviously glad of the company and they talked for a long time. I was happy to listen to their conversation and play with the dogs. The man spoke with the quiet dignity of one reconciled to his fate and I felt a bit sad leaving him there all alone. He handed Pat a small package and asked if he would take it back to London when he was going; some small exchange of love between a lonely father and a lost son. They shook hands and he waved at us as we drove away. On the way home Pat was silent for a long time and didn't snap out of it until we reached Tralee. He pulled to a stop at Dolly Mulchinock's shop. Pat would never return home empty-handed. He knew that the chocolate Crunchy bars with the honeycomb centres were our favourite sweets. Dolly had two large boxes of these on display in her small shop window and, without hesitation, he bought the lot. We feasted on Crunchies for weeks.

Daddy was a safe driver and his speed never exceeded more than forty miles an hour, but his driving career only lasted for three or four years. One incident brought it to a close. Pedal control in Morris Minor cars was a bit tricky, in that you pushed the pedals inwards rather than downwards. Daddy's left hip-joint was getting worse, causing him much pain and restricting his movement on the clutch pedal. This resulted in a misadventure that took place at Ballinagarach bridge. There was a steep incline where the side road joined the main road. He had to yield to an oncoming car and in the process the engine stalled. The hill-start gave him trouble and the car began rolling backwards. In his panic, confusion set in and he lost the plot completely. The car veered to the left, sliding tail first into a deep dyke, leaving the front of the car stuck up in the air. He had to wait there until somebody saw this strange sight of a vertical car with the driver still at the wheel. A crowd gathered to witness his embarrassment but quickly organised a tractor and pulled the car back on the road and brought it home. There was no damage done to the car but Daddy's confidence and pride were badly damaged. He never drove again. So Mammy took up the gauntlet and became the family chauffeur. She mastered the art very quickly and with the help of St. Christopher and Jesus, Mary and St. Joseph she drove into her old age without a mishap. Daddy didn't go with us much from then on, except to Mass and funerals. He would make some excuse and stay at home. Mammy never showed her disappointment and seemed happy to go it alone.

The car opened up a whole new life for her and for us. She would take us to Banna Strand on summer evenings after the day's work was done. Banna strand had nothing to offer but the beach and what a wonderful beach it is. It stretches for miles in either direction and when the tide is out the vast expanse of sand is perfect for playing games of all sorts. And the waves that always seemed to be so big at Banna, made swimming and messing about very exciting. There was one small shop near the entrance that sold ice-cream cones and that was our little treat for the homeward journey. She loved driving and found plenty of excuses for going on trips. She took us to the cinema in Tralee to see *Ben-Hur*. This was the first big film we got to see and it left us awe-struck for weeks. The chariot race in the Circus Maximus would be re-enacted around our yard many times over the following months. The only previous films we had seen were very old black and white cowboy films that Mike Buckley showed in Kilflynn hall. They had now lost their appeal and so each week we kept a 'weather eye' out on the *Kerryman* and tried to coax her to take us to other films. She was a little dubious about many of the titles and didn't want any corrupting influences visited on her innocent boys, so we didn't go very often and only got to see films that had a religious content like *The Song of Bernadette, The Ten Commandments* and *The Greatest Story Ever Told*.

Around that time John B. Keane's plays were gaining popularity and notoriety at the Abbey Theatre in Dublin. He was exposing an aspect of Irish society that 'establishment Ireland' preferred to sweep under the carpet. Holy Catholic Ireland liked things to look right on the surface. An unwed mother and her 'illegitimate' offspring were regarded as a source of scandal in local communities and had to be removed as expeditiously as possible. Those who slipped through the clerical net were often treated as second-class human beings within their own parish. Such is the story of *Sive*, one of John B's earlier works, in which a seventeen year old 'illegitimate' girl is being married off to a farmer of seventy. She loves a young man who is poor, but is being forced into the arranged marriage. Rather than go through with it, she throws herself into the river. Mammy was dying to see this play because she had been introduced to John B in Listowel sometime previously, so when it was being staged in Abbeydorney, she asked us if we would like to see it.

Wow! going to see a real play. This is a first for us and our excitement is building as we put on our 'Sunday clothes'. We run to the car house where Kevin claims the front seat and myself and Tom sit at the back. Mammy is all dressed up too and looking her best. She smells of sweet perfume and is singing her favourite song, Carrig Donn. She fumbles with the car key. She switches on the ignition and the engine goes chug, chug, chug, chug.

Chapter Ten

She keeps trying it again and again till eventually the battery dies. The anti-climax is hard to bear. She sees the disappointment in our faces so she starts to compose a song about it while we sit there in our dejected state. She sings: "Oh! we couldn't go to Sive, because the car we couldn't drive, so we sing this silly song and we hope it won't be long, till the next time". She makes us sing it with her again and again until we get over our disappointment. She is laughing and we are laughing with her. We walk back to the kitchen in a happier mood, disappointment forgotten.

For many years she took us to the annual *Pattern* in Ballyheigue where she would meet with Caseys, Hanlons, Brassils and Dineens. These were all her first cousins, mostly female, from her mother's side of the family. They all loved each other like sisters and when they met they talked and laughed for hours, while we played games on the beach. One year she decided to pay a visit to the home of her favourite cousin, Nell Casey. Dan and myself had the honour of being chosen to go with her. Dan was three and a half at the time and was the apple of her eye, her pride and joy, and I was there to mind him in the car. Nell had never married and lived alone in a house that was as neat and tidy as a doll's house; quite a contrast to what we had at home. Every room was so neatly decorated it left us gobsmacked. There were lots of pretty ornaments on display everywhere. When the house inspection was over she invited us into her parlour for tea and cake and, as was the norm, she took out her best china tea set. It was a really beautiful tea set both in colour and design, but its crowning piece was the little red milk jug with a gold trim around the top. Dan was so fascinated by it that 'he let the side down'; he whispered to Nell if he could drink his tea from the little jug. A moment of stunned amazement was followed by a burst of laughter and without hesitation she emptied out the milk and poured his tea into the jug. Mammy protested but Nell paid no attention; she was excited and delighted to be able to fulfil a child's wish.

Ballybunion was never Mammy's favourite seaside resort but, with the car at her disposal, she always went there on the Feast of the Assumption. This was a Church holy day in August and many people were off work, including the farming community. She went there in the hope of meeting friends that she hadn't seen since last year's visit. As the evening grew late she would walk up one side of the main street and down the other hoping to bump into someone she knew and we would walk with her. On one side of the street there was a cinema advertising all sorts of exciting films, including cowboy films and on the other side, a little further down, was a large ballroom. The music from the ballroom filled the air and for some inexplicable reason it always made me feel nervous and ill-at-ease.

Crowds of people could be seen queuing up to go dancing but there was nobody queuing up to go into the cinema. My child's reasoning could not understand it; watching an exciting cowboy film seemed to me far more interesting than dancing. One could dance around the kitchen floor at home, if they so wished, without having to pay for it. Anyway, what was so good about dancing? It held no appeal for me whatsoever. Some years later it began to dawn on me that there was more to ballroom dancing than just dancing.

Chapter Eleven

School

St. Teresa's National School, Kilflynn, is situated on the high ground above the village and offers a panoramic view of the countryside with the long low ridge of Stacks Mountain in the distance. My school memories date from September 1952 to June '60. Like most rural primary schools back then, it was very basic. At the front of the building there was a wide hallway, which served as a cloakroom and utility room. There was no electricity, but large windows back and front provided ample light even on dull rainy days. The rooms were bitterly cold in winter. The teacher's desk was situated in front of the fire while children at the back of the room got little or no heat. Turf supplied by the parents was the only fuel burned and this was stored in two turf sheds at the rear. A high wall separated the boys' yard from the girls' yard with toilets on either side. The toilets were very primitive. The stink was terrible. The smell of ammonia assailed the nostrils from a far distance and only through dire necessity did we enter. For better, for worse, this was where my formative years were moulded and shaped. Whenever I drive by memories flood back, some are pleasant but many are not. Starting school was not a good experience for me.

On that first morning I can see myself being dragged into Mrs. Roche's room bawling my head off. Mammy is pulling my hand so hard that it's hurting. Her voice is cross and she's telling me to stop crying. Mammy used to say that my "bladder was very near my eyes" and often tried to make me stop crying by slapping me, but Sara and Lizzie would later mollycoddle me and make the world seem right again. Sara's caring voice would say "What's wrong, leanbhin?" But Sara is not with me now and I'm frightened of Mrs. Roche. She tells Mammy to leave and grips my hand firmly as she closes that big green door. I'm calling out for Mammy as I stand there kicking it for a long, long time. I'm vaguely aware of all the other faces looking at me as I desperately try to reach up to the brown wooden knob. I can hear the children's voices chanting some strange words in unison. Mrs. Roche is pretending to ignore the rumpus at the door and is trying to teach her class. The drone of their voices is monotonous and lulling me to sleep. Eventually, I slump to the floor and lie there sobbing and sniffling and fall asleep. When I wake up I find myself sitting beside Billy who's colouring a picture and asks me to help him. The distraction keeps me from crying until Mammy comes to take me home at mid-day. For weeks I rebel and cry my eyes out each morning, but bit by bit I begin to accept it; it's the way of the world but my timid heart keeps fluttering like a captured bird every Monday morning.

Some weeks have gone by and Mammy has asked two older girls, Bridie and Mary Crowley, to come to our front door each morning to accompany me to school. It's a long walk up the Green Road, out past Carmody's house, up Neenan's Hill, on past Buckley's house to Mick Shanahan's cottage. As we're nearing the school my eyes are filling up again and Mary keeps calling me "Cry Baba", but Bridie wipes my eyes and takes my hand. I like Bridie, she has a nice smile and a soft voice. I like holding her hand. I dread having to part from her when we enter the school, but she is very kind and comes into the classroom with me and waits until I am settled in. Mrs. Roche's first lesson is always the same. She points at a picture chart with a long stick and calls out words in Irish: Tarbh, Bó, Lacha, Madra, Coinín and so on to the end. We all repeat the words in a loud response.

At lunch-breaks I watch other children running about as I stand motionless in the same spot by the schoolyard wall; a lone and forlorn figure afraid to move. I feel cold, lost and frightened and wish I could be at home. One day Billy asks me to join in his game of chasing but I am stuck to the ground, my legs won't move. He gives up his game of chasing and stays by my side, panting from all the running he was doing. After a little while he suggests I try to do a short run over to the near wall and says he will run beside me. He counts one, two, three, go! Like O'Flaherty's seagull fledgling leaping from the cliff ledge I'm set free. My legs have taken on a life of their own and are sprinting across the yard. I feel lightheaded and wildly excited. Billy has broken the spell and things begin to look a little brighter from then on.

Billy and I remained good friends all through our boyhood years, even though there was 'bad blood' between both our fathers. Some grievance from the past had remained unresolved and had blighted the relationship between the two families. Daddy was a die-hard in this respect, once a relationship was sullied there was no going back. They never spoke and were at pains to avoid each other. We never knew what lay behind the bitterness. Daddy never attempted to pass on any bad feelings to us but we were aware of the awkwardness of the averted gaze as they passed each other on the road.

One summer evening, in the heat of a game, Billy broke my nose. We were playing hurling in the 'well field'. When reaching in to pick up the ball my head was on a downward motion as Billy's head came upwards. It smashed into my nose and cheekbone with sledge-hammer force. Temporarily dazed, I was unaware of what had happened and how much blood I was losing. The others put me sitting on the ground, pushing my head down between my knees as they had seen somebody do one time, but the blood continued flowing and it terrified me. I had no idea my nose was broken; all I knew was that it hurt. After what seemed like an age the blood stopped and I gingerly made my way home but only as far as the ruins of the old Great House. I hid at the back of its enormous walls nursing my aching head, in the same spot where I had comforted the stray dog the previous summer. The place felt eerily empty. I remained there alone, cold and miserable, like a wounded animal, not knowing what to do with myself. Small drops of blood were still coming and I kept wiping them with my sleeve. I longed for Sara's comforting voice and the warmth of the kitchen. Eventually, Kevin came back to see was I alright and together we sneaked in to the house without a word. I needed sympathy and comforting so badly but was fearful of reaching out for it and suffered my lonely trauma in silence. Nobody's fault really. Mammy's life was often stressful, coping with household chores and young babies, and older children's silly complaints sometimes made her angry. My nose was just slightly misshapen and healed itself of its own accord, until it had a second mishap.

By some strange twist of fate I had a second nose-breaking experience about three years later in the school yard. It brings Lady Bracknell to mind: "To lose one parent may be regarded as a misfortune, to lose both looks like carelessness". Whatever it was, carelessness or impetuosity, it happened again. We used to play a rough and tumble game with a small rubber ball in the school yard during the lunch break. When you got the ball you struck it with your clenched fist in the direction in which your team was playing. I was good at this game and could grab the ball out of the air and drive it long and accurately with my fist. I was marking a classmate, John Shanahan, who was strong and rough. In his determination to win the ball his head smashed into my face with sickening force. I fell to the ground and lay there dazed, with blood pumping from my nose. Donal Leen brought me a rag soaked in cold water and eventually the bleeding stopped, but this time I was aware of the damage done; my nose felt wrong; it felt out of shape. A few lads escorted me into the school but immediately returned to their game. I must have been a sorry sight standing by the cloakroom wall, pale and sniffling, with the piece of rag pressed up against my poor battered nose. The Master and Miss Rice walked by lost in conversation. He was laughing loudly at something she had told him and neither of them noticed the injured creature by the window. The lunch bell rang and I sat in the classroom for a further hour and a half, barely holding it together. That evening Mammy took me to the 'nose doctor' in Tralee and he arranged a bed in the Bon Succour Hospital where he would try to straighten my crooked nose.

The hospital ward is very large. Mammy has left. I feel scared and abandoned with only strange adults all around. Mammy is a close friend of Nurse Regan and has asked her to keep an eye on me. She brings me some sort of a game and a comic. She feels my nose and tells me not to worry and that everything will be okay. But the ward is busy and she has things to do. It's getting late but I cannot sleep. The hours roll by slowly and the wheezing and coughing of the old man in the next bed grow worse as the night wears on. In the stillness of the ward he mutters 'youngfella' a few times but I pretend not to hear; I am too frightened to move. As the hours wear on his breathing goes quiet and the wheezing stops. I have drifted off to sleep but the commotion around his bed wakes me up in the early hours. The old man's bed is now curtained off and nurses are talking in hushed voices. I catch a glimpse of his pallid face as he is wheeled away on a trolley. It's my first time being near a dead person and disturbing images haunt me for the rest of my nights in the hospital. My operation takes place the next day. Because of the previous fracture, Dr. Fitzgerald has to break and reset my nose. He tells Mammy I should keep pressing it to the right side during the coming weeks to keep it straight, but I must have forgotten or neglected to do that because it was never quite right afterwards.

School games followed their usual pattern, conkers in the autumn and marbles in the spring but our ball game was the most popular. But if nobody had brought a ball to school we made up other more boisterous games. When the turf shed in the boys' yard was empty it was used as a 'jailhouse'. It was pitch-black inside when the door was closed. The jailers would target some individuals and drag them kicking and struggling to the jail where they would remain locked up for the duration of the lunch break. One day they targeted Paudie Fuller, who was big and strong, and the jailers had to call for help. Suddenly a big mob was pulling and pushing and trying to subdue him. I joined in but got knocked to the ground skinning my right knee. I watched from a distance as I nursed my wound. His clothes were being torn and sweat covered his face. It was no longer a game and was beginning to turn very nasty as they pushed him into the shed and locked the door. He was shouting dire threats at some of the jailers so nobody would let him out for fear of retaliation. Kevin was a friend of his and risked opening the door as lunch-break came to an end. The Master was unaware of the incident but his appearance in the yard offered protection to the jailers. After school they made a quick exit home and hoped that it would all be forgotten over the weekend.

The Master controlled his small rural school through fear, using a jobber's cane. This was a piece of heavy bamboo stick, about a metre long that cattle jobbers used for prodding and beating cattle at fairs and markets. Corporal punishment was seen as an essential element in maintaining discipline. It was sanctioned by law. Eighteenth century English law permitted the beating of wives by their husbands "if the stick used was not thicker than the thumb". The jobber's cane would just about qualify for such a purpose. Whatever about wives, the beating of children with sticks was seen as a perfectly normal practice. It was mainly the boys who suffered the corporal punishment which, presumably, was often deserved.

As students we became skilled at "reading the day's disaster on his morning face". The creed in most schools back then, and indeed most homes, was that "if you spare the rod, you spoil the child"; a mindset based on the religious belief that children are born with an inclination to evil and therefore adult chastisement was a necessary part of their upbringing. The metaphor of shaping a young sapling to grow straight was often used to justify this belief. Coupled with those beliefs was the reliance on fear of the 'stick' as a means of enforcing learning; a belief that education could be beaten in. As well as the cane various other forms of punishment were the norm; gripping one cheek with the fingers while slapping the other cheek was a favourite of some teachers, but worst of all was being dragged upwards by the ear or by the lock of hair in front of the ear. It was the way things were and we accepted it as the way of the world, but it left its marks. The physical marks disappeared fairly quickly but the psychological marks were more difficult to eradicate.

But there were exceptions. When Mrs. Rice retired her position was filled by her daughter, Miss Rice, who brought a breath of fresh air to our Dickensian school. She was young and vivacious and full of the joys of being a teacher. Her enthusiasm sent little ripples of happiness and surprise throughout the school, even to those who only observed from a distance. She always wore a smile and would engage with children outside of the classroom. Sometimes she could be seen holding a Junior Infant child in her arms as she walked around the playground. As well as bringing a caring and friendly element she added something to the musical life of the school, which was practically non-existent most of the time. She came to our classroom to teach singing once a week and I liked what seemed to be her favourite song: *Will You Come to the Bower*. I envied the pupils in her classroom and wished that I could be there. It was generally known that she preferred to use the carrot rather than the stick to encourage children to learn. But old beliefs die hard. I overheard another mother talking to Mammy one evening about Miss Rice, saying, "it won't work, it's no good, she's too soft on 'em, they'll learn nothing from her". It took another twenty years or so before corporal punishment was banned in Irish schools and a new more enlightened era was ushered in for teachers and parents alike.

The spring water for the school was supplied every morning in two galvanised buckets. This was used for making tea for the teachers and for washing up. One day the Master noticed that some sort of oil had been dropped into the water. The culprit or culprits would not own up so every pupil in the school was brought out into the yard and lined up in descending order, 6th class, 5th class, 4th class, 3rd class, 2nd class, 1st class. With jobber's cane in hand he called on the culprit to come forward and when nobody stepped out he started with the 6th class and sweating and panting he worked his way down the long line of children. The terrified screams of the 1st class children became too much for Mrs. Roche and she engaged in some sort of 'plea bargaining' so that they could be spared the ordeal and taken back into their classroom. He continued the caning in the sure and certain knowledge that the culprit had been punished. After this Pyrrhic victory he disappeared into his classroom leaving everybody in a state of disbelief. Times and attitudes were different then. Not a single parent protested. Parents themselves were often heavy-handed in disciplining their children and many accepted the motto that what you got in school you deserved. Others may have been afraid that their children would be ostracised during their remaining years at school, if they dared to protest.

A lot of time was spent on Gaeilge, religion and handwriting. In the junior classes, English handwriting and Irish handwriting were separate lessons. We practised the old Gaelic script and the ornate English script from a headline the teacher had written on the board. The ink was made in a big jug by mixing the blue powder with water. Our writing copies were distributed, followed by the old-style pens, many of which had damaged nibs and then for half an hour or so we scratched our way between the two blue lines trying to replicate the copperplate script on the blackboard. Blotting the copybook was a punishable offence and so a sheet of blotting paper was essential for safety. In fifth and sixth class we studied geography through the medium of Irish from English wall maps that carried the heading *The British Isles*. Using the Irish language as a medium gave us a good grounding in our native tongue but the subject matter was often lost to us. I could list the names of the towns in every county but couldn't make the connection to their English names. Memorisation was the main teaching strategy and this seemed to have suited me. In my essays I could reproduce large paragraphs that I had memorised but struggled at producing anything original.

I was conscientious about homework and often tormented Mammy with questions while she was working. One evening I brought my arithmetic book to her as she was milking a cross cow. I needed help with the sums. She was trying to explain how to do them when the cow decided to end our conversation. She kicked the bucket, knocking Mammy off the stool and spilling the milk all over the floor. I could hear the anger in her voice as I made a quick escape out the door. The following day I got slapped for not having my sums done. But other children had a harder time. Those with learning difficulties, or without home support, bore the brunt of the teachers' wrath. The education system then did not cater for different learning styles, different teaching methods, individual abilities and differing developmental stages. Child-centred approaches were as yet unheard of, and those who had difficulty with the 'one-size-fits-all' were branded as being slow, dull, or stupid, depending on teachers' sensitivities. Once these labels were tagged on they tended to stay tagged on, undermining self-confidence and self-worth for the rest of the individual's life.

The 'catechism' was a central part of the curriculum. All the catechism questions, of which there were many, had to be memorised and were colour coded for the appropriate age group. We had to know all the easy ones off by heart for making our First Communion. When we had learned all the prayers and had our sins explained to us, our religious knowledge was tested by the diocesan examiner and by the parish priest. Mrs. Roche spent a lot of time in preparing us for our first confession and for receiving the Eucharist. She explained that the host was Jesus and that we had to be careful not to bite it. She had cut out coin-shaped bits of cardboard and would place them on our tongues as a trial run for the big occasion. I looked forward to the big day with nervous anticipation.

Chapter Eleven

Some weeks beforehand Mammy took me to Tralee to get my new clothes. Revington's shop was Tralee's big department store and she was well known there by most of the staff. The first stop was at Eddie Fitzelle's counter for the suit. He chose a navy suit, with short pants and a double-breasted coat. I was happy with that so we moved on to the shirt and tie department. The choice of shirt was straightforward; it had to be white, but the tie presented a dilemma. I took a fancy to a bright red tie, while Mammy tried to convince me that a darker shade might be better, but to please me she went along with my choice. She bought a pair of white ankle socks and then said we should walk across the street for the shoes at Walsh Brothers.

But before she could leave Revingtons she had to visit the china department where her first cousin, Tom Casey, worked. I knew this would take a long time. He was a very close friend and they loved to talk, so I wandered about the store fascinated by the pulley system for sending the money up to the office. With a pull of the rope the little round money container would shoot along the wire to a central booth and in a little while be returned back down with the receipt and correct change. It seemed like a game they enjoyed playing. I would have loved to have had a go at doing it. I made my way back to the china section and eventually we headed for the shoe shop.

I knew exactly what shoes I wanted. I'd seen Jerome McCarthy wearing them at school. Willie Shanahan, who worked at Walsh Brothers, brought out many different styles of shoes, and they all fitted perfectly, but none were the kind I wanted. So my kind, patient, considerate mother traipsed the town with her troublesome son, visiting every shoe shop to try and get what I wanted. But this style of shoe was not to be found. Tired and weary we went back to Willie Shanahan and bought a pair of black shoes that I never liked. She could see that I was disappointed so she asked if I would like to go for something to eat and, without hesitation, I said I'd love to have chips in the Eskimo restaurant. The Eskimo was her favourite restaurant whenever she was in town and, to my inexperienced taste-buds, the chips they served were the most delicious food on the menu.

***

As we got a bit older we would get ourselves ready for school in the mornings while the adults were milking the cows. Kevin was head-chef and would fry the eggs. The heavy cast iron frying-pan was always kept under the range, without removing the grease from the previous day.

When Kevin put it up on the range little teeth marks and scratches were often noticeable on the congealed grease but that never bothered him. He'd fry our eggs and make the tea and wouldn't tolerate any complaints from lesser siblings. Some mornings Mick would be heading off to the creamery just as we were ready to go and he would offer us a lift in the creamery car. The creamery car was just the ordinary wooden horse's cart with the two enormous iron-banded wheels. There was no place to sit so we would stand by the milk churns grasping their handles for dear life as they jigged around on the floor of the car. The wheels made a crunching sound, grinding their way over the stony, pot-holed surface of the Green Road. The horse would saunter along at a steady pace and often we would be overtaken by the Lynch's creamery car that had rubber tyres and a spring suspension. We envied them their comfortable speedy ride, but Mick was blissfully unaware and would entertain us with some ridiculous stories. On mornings when the crows were foraging in great flocks in Jimmy Lynn's meadow he would tell us how he had locked them in there the night before on his way down from Parker's Pub. Other times he would sing a verse or two of his only song that told of the shooting of D. J. Allman by the Black and Tans. When we got to the school he would help us down off the car and tell us to pay attention to our teachers and learn our lessons.

But most days we preferred to walk. We could do it in twenty minutes at a quick pace but there was always the danger of meeting John Weir's bull. We set out one morning without our raincoats because the sun was shining but by the time we got to Neenan's Hill it was bucketing down. In fear of being late we kept on walking at a steady pace until we heard a deep-throated bellow. There in the distance, with the long chain hanging down through the ring on his nose was the big Hereford bull leading the cows in our direction. The road was very narrow. Fear and panic clutched our hearts. We stood for a moment frozen to the ground till Kevin dragged us into the wet grassy ditch, where we lay motionless under the pelting rain. The bull passed by with that terrifying low gurgling sound deep in his throat. We waited for a while and then ran the remainder of the journey to school, now totally soaked. None of the teachers noticed our condition as we sat there, shivering in our wet clothes and longing for three o'clock to come. Time plays a mean game; it slows to a crawl when you want it to go fast and flies away when you want it to go slow. Too shy or too afraid to speak up for ourselves we endured our misery until the clothes dried on our backs. "Children should be seen, not heard" was a phrase we were accustomed to hearing but in this situation we were neither seen nor heard.

Chapter Eleven

One winter's evening a heavy snowfall covered the ground as we walked home from school. It was like a fantasy scene from a fairytale book. The gleaming white snow covered everything. It lay thick on the road and on the fields and weighed heavily on the briars and bushes. It had the perfect consistency for snowballs. Who could resist it, so pure, so white and so soft? There were two groups walking down the Green Road that evening. Our group was slightly ahead of the other group. In our group there were Galvins, Leens, Kennys and Crowleys. In the other group there were Lynchs, Connells, Buckleys and one adult Frank Connell. They had been throwing snowballs at us intermittently, as we walked along and we returned with numerous volleys, but when we got to Garrybawn cross we decided to make a stand and so the battle began. Ammunition was plentiful to left and right. The snowballs began flying at a furious rate. At first we were driving them back but they had the advantage of having an adult on their side and we found ourselves retreating. We had a slight numerical advantage and by making a concerted effort we turned the tide once more and so the battle raged hither and thither for at least an hour. In this excited frenzy reality got blurred and I found myself running on adrenalin alone, in a sort of trance, no longer conscious of my actions. I lost all sense of place and time and sounds and voices became strangely unreal, as if coming from a great distance. My speed-driven actions had become purely instinctual; find some fresh snow, pack the snowball as hard as possible, target an opponent and fire. Then retreat quickly and repeat the operation. It was a weird feeling and later others told of the same surreal experience and I wondered if this was what soldiers feel in the frenzy and mayhem of battle. The fight raged on with no quarter given from either girls or boys. Eventually, Frank Connell, who was now in a lather of sweat and panting hard, got angry and accused Martin Crowley of throwing a stone inside the snowball and this brought the battle of Crotta versus Garrybawn to a close. The result was inconclusive.

***

When I was nine Mr. Sugrue (known to us as Shuckru, from the Gaelic *Síochrú*) joined the staff at Kilflynn school. He was a native speaker from the Dingle peninsula and taught 4th Class in a very confined classroom. The three-teacher school had become a four–teacher school and to accommodate this change the middle classroom had been divided with a wooden partition. Conditions in this room were not conducive to learning but I enjoyed my year in 4th class because I liked Shuckru. Dingle was more than thirty miles distant so he had taken lodgings in Herbert's house near the school but on Monday mornings would come directly from his hometown on his motorbike. One Monday morning he arrived a bit late and was looking the worse for wear.

I'm standing at the classroom door as he comes in. He hands me a sixpence and asks me to go down to Condon's shop for a blue 'jaret', at least that's what it sounded like and I'm too polite to ask what it is or what it might be used for. All the way down to the village I keep repeating jaret, jaret, blue jaret. When I get to the shop Nora Houlihan is busy behind the counter and I mutter out that my teacher wants a jaret. Nora looks puzzled and obviously doesn't have a clue what I want. She tells me I should go back and ask the teacher to write it down. I'm in a tizzy. I can't face back to school without it, whatever it is, so I just keep repeating 'jaret', 'bluejaret'. Denis Condon who runs the post office in a separate compartment of the shop has heard me and comes out to see what's wrong. I explain to him that Mr. Shuckru arrived from Dingle on his motorbike and he wants a bluejaret. Luckily Denis is able to interpret and he gives me a single blue Gillette blade. I run back to school in a happier frame of mind and pray that Denis will keep my shopping encounter a secret so that I wouldn't be the laughing-stock of the school. Shuckru gives us some writing exercises to do and goes across to his lodgings in Herbert's house and comes back a clean shaven man.

Later that year he put me in charge of the library. The library consisted of the bottom shelf in his press and held about thirty books. Shuckru set about raising our standard of reading from comics to books and asked everyone to loan a book to the library if they could. Books were a scarce commodity in our house but my mother had bought a book about Archbishop Stepinac's imprisonment in Hungary and this was my gruesome addition to children's reading in Kilflynn school. Not many books came but Shuckru bought some new books; books like *Coral Island, Kidnapped, The Children of the New Forest, Treasure Island* and a western called *The Dog Crusoe*. I loved those books. I didn't read them, they were way beyond my reading ability, but Mammy was a gifted reader. Night after night she would have us all enraptured, hanging on every word, every tone and every nuance. She knew how to build up the pathos in the sad bits and Daddy would pretend to be fixing the fire to hide the emotion he was feeling. She was equally good at the scary bits and would have us sitting at the edge of our chairs and would sometimes let out a scream just to see us jump. Whole new worlds of strange places and strange people were brought into our kitchen, setting my imagination aflame. Those were wonderful nights that I never wanted to end; no television programme could ever compare.

Chapter Eleven

Canon Sheehan's *Glenanaar* was her all time favourite book but, for us as children, a book about Dick Turpin held a particular fascination. This highwayman would hold up the horse-drawn carriages along the roads of England by concealing a rope across the road and pulling it up tightly when a carriage approached. We had the perfect spot in our farmyard for doing this, so we became highwaymen. There were huge clumps of New Zealand hemp plants, relics of former times, with long fronds reaching six feet high, on either side of our yard. They provided an ideal hiding place for highwaymen. One summer, Daddy had bought a new horse's reins for ploughing, which was about thirty feet long. It was clean and white and never used and just right for our mischief. We would string the rope across the yard and conceal it under grass and weeds and then lay in wait for some unsuspecting family member who was passing by. We would pull the rope tight and attempt to drag them backwards up the yard. On this particular day our uncle, Tom Leen, had come on a visit to discuss something with Mammy, so we were excited to have a new victim to hold up. We set the trap and lay in wait behind the New Zealand hemp plants for him to come walking down the yard. With simultaneous action from either side we sprung the trap and began pulling him backwards, but to our dismay and disappointment, he wasn't amused. He didn't play along with us. Instead he reached into his pocket, took out his pen-knife and cut Daddy's new rope in two and walked on. We stood there dumbstruck not knowing how to react. Our game had gone sour, but worse than that Daddy's new rope was no longer fit for purpose. At first we planned to sneak it back onto the hook in the stable, but Kevin decided we should do the honourable thing and explain what had happened. We could see that Daddy was annoyed but he said nothing. To his credit we often acknowledged that our father never gave out or raised a hand to his children. But Uncle Tom had burst our bubble and Dick Turpin games were no longer the favourite after that incident.

We didn't have many books but there was plenty of reading material in our house. Our kitchen always had a plentiful supply of the same magazines; *Ireland's Own, Our Boys, The Far East, Africa,* and the little red *Messenger.* On Friday the *Kerryman* and *The Farmer's Journal* would be added to this collection and on Sunday two other papers were bought, *The Irish Catholic* and *The Sunday Press. Ireland's Own* was my favourite magazine because the centrefold always had two pages of songs. I could learn the words of all the John McCormack songs as I played them on the gramophone. We had a large box of 78 rpm wax records by John McCormack and a few by Sydney McEwan. Then Mammy started buying new records by Bridie Gallagher. Every time she went to Tralee she would call in to Cable's music shop to see if they had any new releases. She loved the sound of Bridie Gallagher's voice and she loved all those sentimental emigrant songs because the heartbreak of emigration was very familiar to her family. Whatever song she bought I would search for it in the stack of *Ireland's Own* on the window ledge and if I didn't find it there I had another source; T.J. Hannon at school. T.J. was in my class and he also was a collector of songs, so we regularly swapped. I would cut out a few from the *Ireland's Own* and see what he had to trade. He had a good voice and I learned many tunes from him. We were rarely taught any English songs at school and the few Gaelic songs Shuckru tried to teach us were very difficult and beyond our understanding.

My library job entailed staying behind after school two evenings a week to record books going out or being returned. Thankfully trade was generally very slack and I was not under much pressure. Shuckru would be there sometimes and would talk to me about ordinary everyday things. This was a new aspect of a teacher; I thought they only spoke about school stuff. He would refer to me by my first name, which made me feel very valued. But Shuckru could be tough and severe too. As I fixed up the library books one girl would always be there in the room; left behind, a solitary figure sitting at her desk, trying to complete some writing exercise and crying into her copybook. Her name was Teresa. Whenever Shuckru went out she would ask me how to spell some word or to explain what something meant. Teresa was always late for school in the mornings and Shuckru would insist she made up the lost time by completing the exercise he assigned her. The stillness of the empty classroom always felt strange and foreboding and I used to feel sorry for her sitting there on her own in the back desk. She lived down the road from our house and I would sometimes see her trudging home alone, long after I had dumped my school bag on the kitchen window and gone out to play hurling in the front field. On my library evenings I would wait for her but I was always impatient at how slowly she walked. She never seemed to have any energy and I was always in a hurry to get home to the hurling. Nobody knew then that Teresa was suffering from a terminal disease. She died a few years later. At huge crowd attended her funeral. Shuckru stood by her graveside and wept uncontrollably.

During my year in 4th class, Phil Cahill, a dancing master of some renown, offered dancing classes after school to those who were willing to pay. "Boys dancing! no way, no way, Mammy", was my immediate response but somehow she persuaded Tom and myself to join up. Traditional music and dancing were highly prized in her own family. Our grandmother, Hannah Hanlon, was a noted concertina player, who played at weddings and 'American wakes' throughout the neighbourhood. Mammy was glad of the opportunity to bring a bit of culture to her children and found the money somewhere for us to learn Irish dancing. There we were, two boys, learning to do a hop jig and a hornpipe with all the girls, while Phil Cahill whistled the melody to keep us in time. If we didn't have our lightweight 'Sunday shoes' we could dance in our socks. I picked up the steps quickly and became good at it in a short space of time and was beginning to shed any regrets about attending. But this was to be short lived. One day out of the blue the Master summoned me to his classroom. It was wintertime and I was wearing the heavy, hobnailed boots that Daddy insisted I wear. I hated those boots; they were so heavy and clumsy, but Daddy saw them as good value for money because they lasted a long time.

Anxiously I approach the Master's desk wondering what I have done wrong. All the sixth class boys and girls have stopped whatever they were doing and are staring at me. He asks me if I attend Mr Cahill's dancing classes and what dance I'm learning. I tell him it's called a hop-jig. He then orders me to stand over to one side and show his class what I have learned. Surprised and flustered I say I can't do it in the heavy boots, but I know refusal isn't an option so I start to undo my laces knowing I'll be alright barefoot, but in a harsh and commanding tone he orders me to "Just do it". With rising humiliation and embarrassment I attempt to go 1-2-3 and 1-2-3 in my clumsy hobnailed boots. I know how I look: like Aesop's performing donkey; a plough-horse in a show jumping arena. I know they are all sniggering behind their hands, but too afraid to laugh out loud. Without making any comment he tells me to return to my classroom. I slouch to the door, studiously avoiding eye contact with the others. As I close the door behind me I hear their laughter ringing in my ears. I pause in the hallway for a while before entering my own classroom. I sit at my desk and try hard to focus on the Geography lesson. I'm glad to be back in the safe haven of Shuckru's room.

At three o'clock, I'm happy to wait behind with Teresa to avoid the smart remarks that I know the older boys will make. On our way home I tell her what happened and as we talk about it I begin to feel a bit better. Her comments are strangely consoling and I'm glad of her company. We are both feeling sorry for ourselves and gain some comfort from our shared unhappiness. I'm no longer in a hurry home to the hurling and am quite happy to stroll along at her pace. We rest at Buckley's front wall and Willie Buckley, from 3rd class, tells us the story of a cowboy film he *watched the week before. Willie's enthusiasm about the Wild West is infectious and strikes a chord in my imagination. His description of the film cheers us up and his demonstration of the Indian attack makes us laugh. We continue on our way trying to hold on to the happy distraction Willie had provided for us. From Neenan's Hill I can see the remains of the old Great House silhouetted against the evening sky. Beside it our house with its yellow painted walls is partially hidden by the trees. Dark smoke is rising from the chimney and making interesting shapes in the sky. It is getting late and I know Mammy will start to worry. I ask Teresa to hurry on but she can't walk fast so I carry her bag until we come to the end of the road where our paths diverge.*

When I got home I told Mammy I was giving up the dance lessons. I didn't explain why; I just said I didn't want to do it anymore. She looked a bit disappointed but said okay, if that's what I wanted. I felt bad. My dancing career was over before it had begun, but Tom continued with the dancing classes and became the 'twinkle-toes' of the family. Whenever we had visitors or family gatherings Tom would be called on to perform and to the plaudits of everyone he grew in confidence and self-esteem. But Miss Redican was waiting around the corner to take the wind out of his sails.

Shuckru left and was replaced by Miss Redican, a low-sized, slender woman of mature years who never smiled. She was a strong believer in 'mental arithmetic' as a way of sharpening our computation skills. She would have us stand around by the walls of her room and her stern gaze would move from face to face with a mental arithmetic problem for each child. The old money of pounds, shillings and pence, was her favourite subject: *"What would you get if you added one and nine-pence to two and six-pence? Take four-pence ha'penny from a shilling. How many half-crowns in two pounds ten"*? When Tom's group came into fourth class they were all being screened in this way while we, the 5th class, sat there watching them. Myself and Kevin were considered 'good' at our mental arithmetic so her expectations of Tom were high, but alas! He failed the acid test, he didn't measure up. As he struggled with some incongruous mathematical calculation she moved a bit closer and looking him straight in the face she asked: "Are you bad"? Her misuse of the word 'bad' was commonplace in schools. There were two categories 'good' and 'bad' Being 'bad' at lessons put a brand on a child for life. Tom didn't reply, how does a child respond to such a question, but I could see the hurt in his face as he was put standing to one side. She moved along the line weeding out other 'bad' nine-year olds.

Chapter Eleven

Some years earlier Tom had made his debut on stage for poetry recitation. When he was in 1st class, the Abbeydorney community had decided to put on a variety concert in the local hall and Mrs. Roche, being from Abbeydorney village, entered three of her pupils in the concert: Joan O'Brien, reciting *Flo's Letter*, Noel Fuller, reciting *The Man in the Moon* and Tom reciting *Somebody's Mother*. For weeks in advance he was coached on his delivery of this poem, both at school and at home. By the end of this time we were all sick of hearing it and had it off by heart. On the night of the concert Tom gave a powerful rendition. His child's voice was loud and clear. Without a pause or a single mistake he held the attention of everyone present.

> *The woman was old and feeble and grey,*
> *And bent with the chill of the winter's day.*
> *The street was white with the recent snow,*
> *And the woman's feet were weary and slow.*
> *She stood at the crossing and waited long*
> *Alone, uncared-for amid the throng.*
> *Down the street with laughter and shout,*
> *Glad of the freedom of school let-out*
> *Came the boys like a flock of sheep,*
> *Hailing the snow piled high and deep.*
> *Passed the woman so old and grey*
> *Hastened the children on their way.*
> *None offered a helping hand to her,*
> *So weak, so timid, afraid to stir.........*

To make a long story short one boy came to the lady's aid and helped her across the street and then went back to play with his friends and that night the lady remembered the boy in her prayers. Not a sound was to be heard as Tom finished his poem. The diminutive figure on the distant stage had sent an emotional surge through the crowded hall. Many women were sniffling and wiping their eyes. Some reached over to congratulate Mammy on having a son with such talent. He got a rousing round of applause and was showered with praise as he made his way back to where we were sitting. He had found fame at seven and for years to come would be expected to deliver his 'party piece' at family gatherings, while his talentless siblings looked on in envy.

One day in 5th class, Fr. O'Brien, the local curate, dropped in unannounced to Miss Redican's room at religion time. Everything had to give way for him to examine us in our catechism and prayers, as this was our Confirmation year. He was small and extremely thin and like Miss

Redican he seemed to have lost the ability to smile or laugh. All of us in the 5th class were standing around the walls in the usual formation as he checked our ability to rattle off prayers. He would select a prayer and pointing a bony finger would ask the pupil to say the prayer. The list was long: *The Our Father, the Hail Mary, the Glory be, the Act of Contrition, the Grace before Meals, the Grace after Meals, the Morning Offering, the Night Offering, the prayer to the Angel Guardian, the Hail Holy Queen, the Prayer before Confession, the Prayer after Confession, the Prayer before Communion, the Prayer after Communion, the Confiteor* and so on. I thought I was lucky on that day when I got the Hail Mary.

Standing directly in front of me he points his bony finger and says "You Boy, the Hail Mary". I rattle it off at great speed and finish with a loud Amen. To my surprise he stares angrily at me and says, "Say it again, Boy". I repeat my performance at a slower pace, but to my surprise and nervous anxiety he repeats, "Say it again". By now Holy Mary is becoming my least favourite person as he brings me to the front of the class to say it again and he tells everybody to observe closely. When I'm finished he looks at Miss Redican and says, "Did you notice that" and to the children, "Did you notice it". Miss Redican nods but her face says she hasn't a clue what he's on about. Neither of course do the class and neither do I. Then all is revealed. Pointing at me again (he liked to point) he declares to all, that "This boy doesn't bow his head when he speaks the name Jesus". So he asks all the class to watch my head as I say the Hail Mary one more time. Embarrassment showing on my face, I slow down as I come to the middle line: "Blessed is the fruit of thy womb, Jesus" so as not to miss the target. I bow my head so much that my chin touches my rib cage. When I finish he admonishes me to remember that and orders me back to my place by the wall. With a deep sense of being wronged welling up inside, I stare back at him thinking vengeful thoughts, but what's the use, so I try to push it to the back of my mind as I had done many times before. For him the job is complete; he has done his priestly work and served God well. Without a word, he disappears out the door.

Some weeks later I experience a second humiliation. We are going out for 'big' lunch (12.30 – 1.00) when The Master asks me to take responsibility for ringing the bell at 1.00. I feel honoured to be chosen for the task and I put the bell on the mantelpiece beside the big clock so that I can run in and check it from time to time. I gobble down the two slices of my mother's brown bread along with the half pint of milk in the lemonade bottle and get ready for the match. This was the highlight of the school day. I'm defending at one end and striking the ball, with clenched fist, out to the middle of the yard. Dan Shanahan and Arthur Spring, both equally tall, are competing against each other for the dropping ball. I'm on Dan's team and we desperately want to win today's game. The free-for-all match flows back and forth at tremendous pace but the score remains level throughout. I'm playing my heart out. I'm totally engrossed in the game and loving every minute of it. Then I hear somebody saying, "We're getting a very long lunch-break today" and "Oh my God"! Shock! Horror! I remember. I dash into the room and grab the bell. The clock is reading 1.15, fifteen minutes over the time. I'm ringing the bell furiously up and down the yard when I see the Master approaching. He comes straight towards me and without a word of warning gives me the full force of his open hand across the side of my head. The force of it knocks me to the ground. Everybody stands watching as I struggle to my feet, dazed and embarrassed. He snatches the bell from my hand and begins ringing it. I stand there, confused and angry with tears welling up in my eyes. My ear has taken a sickening blow and my dignity a humiliating battering. That sense of being wronged stays with me for a long time and I vow that one day I will challenge him on it. When I tell Mammy about it that evening she is upset and feels sorry for me but says I should do the Christian thing of "forgive and forget". I try to say that I can't forget it but she has no time to debate the point, she has work to do.

***

I made my Confirmation in 5th class. The 5th and 6th classes in Kilflynn school were grouped together with the 5th and 6th classes in Abbeydorney school to make it worthwhile for the bishop to come. So Kevin and I would make our Confirmation together. We had spent months studying the catechism questions in order to be ready for the parish priest when he came to examine us. Mammy had spent endless hours helping me memorise the answers and I could recite all of them without hesitation. Some of the questions had very long answers but on the day of examination I had no problem. My question was: *Why does the bishop give those he confirms a stroke on the cheek?* I sang out the answer with confidence: *The bishop gives those he confirms a stroke on the cheek to remind them that they are soldiers of Christ and that for His sake they must be ready to suffer anything, even death itself, rather than deny the Faith.* The phrase about 'stroke on the cheek' brought back an unpleasant memory and gave rise to a negative feeling that I was finding hard to overcome. Shortly, I would be receiving the sacrament of Confirmation and I should be in the state of grace. Thankfully, on the day of Confirmation, the bishop's stroke on the cheek was just a gentle touch.

But Confirmation was also a 'rite of passage' to manhood and this was marked by the long pants. Prior to Confirmation it was the fashion for boys to wear short trousers, held up with elastic suspenders over the shoulders. In winter time we wore long woollen socks but our knees were always exposed to the elements. Confirmation was to change all that. Well in advance Mammy took us to Jim Lynch's tailor shop in Lixnaw where we got measured for our new suits with long pants. On the day of Confirmation I felt a bit strange in this new rigout but Mammy said I looked very grown up. Unknown to anybody Sara had bought watches for us. She had selected them from an advertisement in *The Sunday Press* by a Dublin jeweller and they had arrived in the nick of time. I loved that watch. I wore it with pride on the day of Confirmation and all through my teenage years and when it finally got tired of ticking I put it in my treasure box as a cherished memento of Sara's unstinting love.

We had got a thorough grinding on all the Catechism questions for Confirmation. We had studied the Ten Commandments in detail with the exception of the sixth which was glossed over. References to human sexuality were studiously avoided. As puberty approached the 'dirty talk' began to set in. In the absence of any kind of sex education either in the home or in the school, the schoolyard became the source of all knowledge relating to human sexuality and much misinformation was being spread about. For months my reaction was total disbelief. I was growing up on a farm and was quite familiar with animal reproduction but my mind could not accept that anything so gross applied to humans. Animals were animals but we were human beings made in the image and likeness of God. We were told that babies were a gift from God and we didn't question the matter further. The two youngest additions to our family, Diarmuid and Pat, had arrived on the scene unheralded. Mammy went to the hospital, remained there for a few days and by some miracle came home with a new baby. She had often explained to us how a fertilized turkey egg would develop into a fledgling in the shell. But a fertilized human ovum and how this fertilization came about was in the realms of the unspeakable and the unthinkable. The whole idea was repulsive to my mind and I refused to believe it. As there was no knowledge of correct terminology for body parts, all sorts of crude words and vulgar jokes were becoming the norm in the classroom, when the teacher wasn't present. It was shock treatment to my innocent mind and it made everything sound dirty and ugly. But summer was on the way and I could put all that behind me. I could switch off school and all its pressures for seven and a half weeks and enjoy the freedom and careless abandon of Crotta.

Chapter Eleven

Time flies by. The summer holidays have come and gone and I'm in my final year at school. I'm sharing a double desk with Billy at the back of the class in the Master's room. We have chosen the back desk as it seems to offer some degree of sanctuary. The desk carries the initials of many former occupants who had wanted to leave some memorial mark behind. The little white inkwell is still in place but is now unused. We have gone modern and are using fountain pens. The Master wouldn't allow the use of biro because it would spoil our handwriting. A bright beam of sunlight is streaming in the window highlighting the curling smoke from his cigarette as he teaches a maths lesson at the blackboard. We are revising 'subtraction of money' sums; long convoluted sums in pounds, shillings, pence and farthings. Billy has always struggled with maths and has no confidence in himself. The Master explains what to do with a loose farthing or ha'penny on the top line of the sum; you "Bring it down". Billy seems to have latched on to this phrase; it sounds knowledgeable so when the Master asks him what to do with a different, unrelated sum Billy, not having a clue, chances his arm and says "Bring it down". The Master looks at him for a moment and then, without a word, lifts the heavy wooden blackboard off the easel and brings it to our desk. He places it on top of our books and says "What do you want me to do now?" Feeling embarrassed Billy attempts to laugh as we all do. Even though it's done in good humour and meant to amuse, our laughs are cautious and muted. Billy nervously stutters "Níl fhios agam (I don't know)" and waits expectantly. There's a long pause, but the moment of tension passes. The Master takes one more puff of his cigarette, remains at our desk and patiently gives Billy a step by step explanation of how the sum is done.

Sometime in October of that year, Gerald Wynn, a native of Kilflynn, had returned from England where he had practised the martial arts of self-defence. When we heard at school that Mr. Wynn was putting on a judo demonstration in the community hall on Friday night, we asked Mammy to take us there. On the night, Gerald Wynn explained what to do if somebody was attacking you. His nephew, Larry White, was on stage with him and very skilfully he caught Larry by the lapels of his jacket and falling backwards pushed his right leg against Larry's stomach and propelled him up and over onto the stage floor with a loud thud. Various other tricks of self-defence were also explained. To me, who was frightened of my own shadow, this was exciting stuff that offered some hope for future encounters with bullies.

The following day, I decide to try it out on my brother Dan who is six. I want to practise the up and over body throw so I make up a soft pile of hay at the back of the hayshed on which he would land. Dan is a willing accomplice and I assure him it's not going to hurt. I show him exactly how it's going to work. We position ourselves strategically in front of the bed of hay. I catch both his hands and as I fall backwards I put my right foot against his stomach. Up and over he goes, flying through the air, just as Mammy comes around the corner. She screams at me and charges in with both hands flailing. Once again my ear receives a numbing blow that's quickly followed by a second and a third. She is still shouting at me, "You could have broken his neck", as she takes Dan by the hand and walks toward the house.

After a moment of stunned confusion I take flight. Like a scalded cat I run back the field as fast as my legs can carry me, not knowing where I'm headed. Exhausted, I stop by the Promsey spring well at the far end of the wood field. I kneel beside the out-flowing stream and drink, animal-like, from its crystal clear water. It helps to cool down my burning face. I continue on to the corner of the field where a wooden bridge leads on to the ruins of the old Crotta creamery. I sit on the crumbling wall in a confused state. I'm bent on running away from home and causing my mother as much anxiety and worry as possible. But where should I go?

I head back the road towards Moll Grady's shop, but I have no money and anyway I don't want to talk to anyone, so I climb over a gate into Whelan's field and sit there for a long time, feeling lost and lonely. I count the crows as they go flying by to the tall tree at the end of the field. I crouch low into the ditch when I hear voices going by on the road and hope that nobody has seen me enter the field. Small birds are singing in the bushes nearby but their chirping brings little joy. To amuse myself I pick stalks of grass and twist them into different shapes that look like birds with long legs. A reddish sun begins to sink slowly on a far off hill. Many hours have gone by and I'm feeling cold and hungry so I sneak my way back home and hide behind the hedge that surrounds our backyard. I stand there shivering not knowing what to do. I try to peep over the hedge from time to time to see if anybody has missed me or is worrying about me. Mammy must have been keeping an eye out because she sees me and in a jeering voice she says, "We see you, we see you", followed by a curt command to come in and have my supper. Cowed and embarrassed I slink in and sit at the table. Nothing is said of my judo tackle on Dan, conversation goes on as normal. My dramatic gesture of defiance didn't even register on the household Richter scale.

Chapter Eleven

I must have been showing some academic promise in 6th class because the Master advised Mammy that I should do entrance exams for various colleges that offered scholarships. He selected five of us out of the class to come to the school on Saturday mornings to do exam preparation. He had bundles of exam papers dating back many years and we worked our way through these over the coming months. We became very familiar with the type of questions that were set and how they were posed; it was almost like seeing the paper beforehand when we came to do the exams. He was proud of the reputation Kilflynn school had gained through the number of scholarships its students had acquired over the years. Some years later, much to his annoyance, the practice of Saturday teaching was banned by the Department of Education. The exams would test our knowledge in three subjects; English, Irish and Arithmetic. The Saturday morning classes were enjoyable and very different from normal school. Mammy showed her appreciation of this extra tuition by killing the last of the goslings which by then was a fine fat goose. She took the wrapped up box to the school on the pretext of asking what requisites I would need to take with me when doing the exams. I sat three different entrance exams in all, but didn't pull off any coveted prize.

The first exam was for De La Salle College, Waterford. A central venue in Cork city was arranged for this entrance exam. Driving to Cork was too much of a challenge for Mammy so she enlisted the help of her fearless, mad-hatter cousin, Kitty Leen, in Lixnaw. Kitty was an outgoing young woman in her early twenties and was delighted to oblige. Mammy and I set off for Lixnaw at an ungodly hour of the morning. Kitty was waiting and sat in behind the wheel. A light rain was falling, which made one of the wipers squeak its way back and forth across the windscreen. Kitty asked jokingly if there was a cat hiding in the car. As we passed Abbeydorney she cast off her high heel shoes and drove the rest of the way in her bare feet, at a much faster speed than we were normally used to. Mammy's right leg was pressing hard on an imaginary brake pedal on the passenger side and her increasing discomfort was noticeable. When Kitty asked her what was wrong she whispered that "Our Morris Minor doesn't like to go beyond forty miles an hour". It took Kitty some time to adjust to driving so slowly and Mammy had to repeat her request on a number of occasions. But the journey passed quickly; Kitty's high spirits and funny stories kept us amused all the way to Cork. She kept putting funny names on the towns and villages we passed through which had Mammy in uncontrollable fits of laughter. Cork city was new to her and it took some time to find the venue in Washington Street.

I was tired and frazzled and ready to fall asleep as I sat down to do the test. The huge examination hall was cold and strange and full to capacity of twelve year old boys. I was allocated a desk and sat there feeling very unsure of myself. Mammy and Kitty went shopping but first they dropped in to a nearby church to light a candle for my success. When the exam was over they took me to lunch in a fancy restaurant. It had been a long time since breakfast and I could hear my stomach rumbling. The chips tasted even better than the Eskimo chips in Tralee.

A week later a letter came to the school saying that I had won a small scholarship of twenty pounds a year for getting full marks in the Arithmetic test. The irony of it was that Arithmetic was my weak subject. Kevin could solve in his head what I had to do on paper, but I was so well groomed by the Master that the maths formulae in my head did the work for me. But De La Salle College Waterford sounded like a foreign country to us and it was very far away from Crotta, so instead it was decided, wisely, that I should go to St. Brendan's in Killarney.

Chapter Twelve

Catholic Faith

Mammy had strong religious convictions and a great devotion to Our Lady. Holy Mary was her first port of call when disaster struck or was pending. I think she believed it was more respectful to use an intermediary than go directly to the Boss with petitions. And her petitions were numerous. She prayed for everyone and everything. The trimmings of the rosary had become longer than the rosary itself. Ten o'clock was always rosary time in our house. Whatever we were doing would have to come to a sudden end when Mammy reached for the rosary beads. An assortment of rosary beads hung by the side of the front window on a crude four-inch nail that had been hammered into the three hundred-year-old carved casement. Our rosary was a long and complicated ritual in which we all played a part, but Mammy carried the show. How she could remember, unaided, all those prayers and the endless litanies that followed was a mystery, while we just repeated a monotonous refrain, *prayfruss*. At the start of each decade Mammy would call a name and like a greyhound let loose that person would tear into the Our Father, the ten Hail Marys and the Glory Be. But Daddy was fastest of all, he only hit the words here and there at the beginning, middle and end, the rest was a blur. The rosary was said kneeling on the concrete floor, with elbows on the *súgán* chairs facing towards the back of the kitchen. Searching in the *súgán* ropes of the chairs we would retrieve the missing spoons, forks, coins and other small objects that the younger clutch, Dan, Diarmuid and Pat, were forever hiding there.

But there was one greater source of amusement at rosary time and they were the mice. In spite of the drone of prayerful voices they knew instinctively that movement had ceased and this was their cue to venture forth from underneath the dresser and scurry hither and thither looking for scraps. In our English Reader at school there was a poem called Mother Mouse, which we had to memorise, and during the rosary its words would pop into my head making me feel sympathetic towards the poor little mother mouse:

Chapter Twelve

Beneath the kitchen in our house, there lives a busy mother mouse. Her family cannot be fed, till we have all gone up to bed. Poor Mother Mouse will often state that ten o'clock is much too late for little mice to have their tea. So she must find another house with ways more suited to a mouse.

When Mammy saw that the mice were winning the attention stakes she would let fly with the nearest missile at hand, usually one of her slippers. At other times she had to compete with the hatching hen for our attention. In springtime a hatching hen, or maybe two, would become part of our household and would occupy a corner of the kitchen as they hatched out eggs in hay-filled wooden boxes. Invariably during the rosary they would hop out of their boxes and parade around the kitchen, cluck-clucking at everything they saw and sometimes leaving messy deposits on the concrete floor before returning to the job of hatching. While all these distractions helped pass the time, the lack of concentration sometimes made the rosary a bit longer. The person saying the decade would overshoot onto eleven or twelve Hail Marys and Mammy would have to call out "Glory Be".

She had an old tattered book entitled *Around the Boree Log* in which there was a poem called *The Trimmings of the Rosary* that told the story of extended Irish rosaries, but ours was a match for any of them. Any disaster, or worthy cause, at home or abroad, merited a mention and not just once but on a continuous basis so that the trimmings grew and grew until Divine intervention was seen to have taken place. Then there were the bereavements and the illnesses that friends and relations had experienced; all were remembered and interceded for, over an appropriate time period. The conversion of Russia, however, was permanently on the list. What good Catholic home anywhere in Ireland didn't fear and loathe the Communists who would make atheists out of all of us if they got half a chance. Hadn't they invaded Hungary and mercilessly put down their bid for freedom, leaving the country in a state of turmoil. So that was added to the trimmings; we'd say three Hail Marys for the suffering people of Hungary. When I heard that Ireland had taken in Hungarian refugees and that the holding camp in Limerick was looking for families to take children, I urged Mammy to apply. The idea of having another hurling playmate sounded great and for a while I was hopeful that it might happen, but common-sense prevailed; she had enough on her hands.

And then there were my own private trimmings for the poor little pheasant fledglings. When I was ten Ned Neenan was hired to cut the hay. Mechanical progress had replaced the horse-drawn mowing machine with tractor and mowing bar. This was much quicker and Ned Neenan and his little grey tractor became very popular. Ned was a good humoured man who was a little deaf and he always spoke with a shouting voice over the whirring noise of the engine. He was a child at heart and a little bit daft and added much excitement to our lives. One evening as he was finishing a meadow he spotted a hen pheasant running out of the tall grass. He stopped the tractor and ran towards the spot and having searched around for a while he found the nest. In it were two little fledglings as yet unable to run. He brought them to me as I stood by the gate and said if I put them in with the young chickens that they would survive in the farmyard. I took them home and made a nest for them at the side of the hayshed, but they kept on calling out faintly for their mother. So late that evening, when Ned had gone, I walked back to the meadow carrying the two little orphans in the nest I had made for them. I placed them on the grassy margin close to where Ned had found them. As I walked away I could hear their faint cry and I had a sinking feeling that their mother would never find them and they would die. So that night at the rosary I silently added them to the prayer list. I visited the meadow the following day. The nest was there but they were gone. I prayed that the mother had come back and taken them to safety but each night for many weeks I silently added them to the trimmings of the rosary.

As a three year old, Dan had the honour of bringing the rosary to a close with an incantation to 'Lady Lourdes'. This was a prayer to Our Lady of Lourdes, to mark the anniversary of the apparition of the Virgin Mary in Lourdes in 1958. It was at Sara's request, as Dan was the apple of her eye and also because she had a special devotion to Our Lady. In the late 1920s, after the birth of her second stillborn child, Sara went with the diocesan pilgrimage to Lourdes. She hoped, perhaps, that the Virgin Mary would intercede on her behalf and that a miracle baby would be granted to her. She brought back a statue, adorned with luminous beads, which remained on her dressing table all through her life. As a sad remembrance of the past, it stands there still, covered in dust and green mould, facing the bed where Sara slept. Each night as the rosary came to an end she would call on Dan to say "Our Lady of Lourdes" three times and we would all respond saying "Prayfruss". Most nights Dan would have fallen asleep and we would have to wait for him to wake up and then in his sleepy, child's voice he would mutter "Lady Lourdes, Lady Lourdes, Lady Lourdes".

***

Chapter Twelve

The Stations, as they were known, were occasions of great importance for the host family. These were house Masses that operated on a rota basis within the parish and involved every household in turn. All the neighbours for miles around would attend and so when our turn came there was much work to be done in getting the house and its surroundings ready. Mick did the white-washing, brushing on liberal coats of lime wash onto the old stone walls. A high curved wall ran from the southern gable of the old Great House and joined onto the orchard walls at the entrance gate of the lower yard. All the walls inside this enclosure would have to get a fresh coat of white-wash. The ivy had colonised the walls creating arched patterns along the lower half of the wall where it was rooted. Mick would white-wash these in-between patches giving the whole area an exotic appearance of white and green. The backyard would be swept of its mud and slush and be given a general tidy-up as most neighbours would enter the house via the backdoor. The parish priest would, of course, come to the seldom-used front entrance, so this area had to be given special attention. Well in advance Mammy had started the wall-papering and painting. She had a flair for colour coordination and interior decorating and seemed to have an endless reserve of energy. She focused her attention on the two 'important' rooms, the kitchen and the parlour. Both rooms had been built in a grand scale and presented a large wall area to be covered. There was the added problem of rising damp. Even though the walls were four feet thick dampness was particularly bad in places so she attempted to dry-line it with sheets of plastic she had purchased from a hardware shop. After weeks of hard work, often running late into the night, everything was finished and the house was looking spick and span for the 'big day'

On the morning of the Stations Sara prepared the parlour for the parish priest and for some self-confident parishioners who were willing to join him for breakfast. She released the best cutlery and china from their long term of confinement in the locked sideboard and set up the table with all sorts of little fancy pieces. She had rolled the butter into butter-balls, using the two specially carved wooden bats with great skill. She placed the jam in little dainty jam-holders that sat on either side of the silver toast-holder. As children we gazed in delight and awe at our first image of sophisticated dining and felt the seeds of a priestly vocation beginning to take root. The neighbours would all enjoy tea and cake and sandwiches in the kitchen in a more rough-and-ready fashion, after the Mass was over. The sturdy kitchen table had received a higher calling. It had been raised up from the floor to make it the correct height for an altar. Mammy covered it with a white tablecloth and placed two tall candles, two vases of flowers and a crucifix on top. This was all very new and exciting for us but because it was a religious occasion we had to curb our enthusiasm.

As Mass time approached neighbours began to drift in and take up positions on chairs and benches in front of our altar. When Fr. Lenihan arrived he was ushered down to the parlour where he put on his vestments and then heard confessions of those who were brave enough or sinless enough to kneel in front of him in broad daylight. Then the Mass began. This was pre-Vatican II when the priest kept his back to the congregation and spent most of the time in whispered Latin prayers. Fr. Lenihan was missing some front teeth and as he prayed his breath made a whistling sound as it escaped his mouth, much like the hissing of a bicycle tyre that had run over thorns. As we all knelt there, looking at Fr. Lenihan drooped over the altar, Mikie Kelly whispered to Mammy: "I think he's punctured". This drew a spontaneous yelp of laughter which she tried to suppress with a muffled cough. After a temporary silence the Mass continued without further incident. When it was over the people made their monetary contributions to a sombre-faced priest who then retired to the parlour for his fancy breakfast. In truth Fr Lenihan was a simple down-to-earth man who placed little value on the pomp and deference being paid to him. The crowd in the kitchen tucked in to whatever was going and amidst the buzz Sara kept filling up their mugs and cups from an enormous teapot. After a few hours they all had drifted away and things came back to normal.

***

Even though we were in Kilflynn parish we always went to Lixnaw Mass. There were three priests in Lixnaw parish, the parish priest, Fr. Browne and his two curates, Fr. Hickey and Fr. O'Keeffe. Fr. Browne was very old and didn't say Mass very often, so it was usually one of the curates on Sunday mornings. When Fr. Hickey left he was replaced by Fr. Scanlon who was noted for his very long, impassioned sermons. As children we liked to see Fr. O'Keeffe coming out of the sacristy, because his Mass was the shortest. He was always in a hurry and seemed impatient and irritable and on one particular Sunday he lost his temper during his sermon. Some men at the back were not paying attention and he reprimanded them in a loud voice saying "Some members of this congregation are colossally ignorant". 'Colossally' was a new and difficult word being brought into the Lixnaw vernacular. Bouts of coughing broke out from the back of the church followed by an eerie silence. We hoped Daddy was not one of the guilty ones. He always stayed at the back door with the other men, while Mammy marched her brood up front. Before Mass, it was the custom for men to stand at the railings outside the church until the second bell was rung. Then they would approach the church door and pay one penny to Moss McElligott or Tom Lovett, whose job it was to collect the donations. But times were hard and some people would only put a ha'penny in the box. When the Easter dues were read out they ranged from Dr. Nora's five pounds at the top of the list down to somebody's one and six pence at the bottom of the list. Mass was always well attended and most people would receive communion. When distributing communion Fr. O'Keeffe could only operate from left to right. So when he worked his way to the end of the long marble altar rail he would have to dash across to the other side to begin again, with the altar-boy racing behind him. Everything was done at speed during his Mass. He didn't have patience to say the full Latin incantation of *Corpus Domini Nostri Jesu Christi...* as he distributed the host, so each recipient received just a muttered version.

The various rituals associated with Catholic worship provided a welcome excitement and structure to our lives. My favourite ceremony was Benediction. The smoke from the burning incense wafting around the church had an intoxicating effect and the melodious sound of the nun's choir coming from the adjoining oratory induced a state of tranquillity that was almost hypnotic. Their rendition of *O Salutaris*, the *Tantum Ergo* and other hymns was truly beautiful. Fr. Scanlon would walk about the Sanctuary with the golden-spiked monstrance held high, intoning Latin verses to which the nun's would respond; their disembodied voices reaching us through the open oratory door. In those days no female footfall was permitted within the Sanctuary and no female curls would be visible in the church for fear of distracting the male faithful from their prayers. Inside the main door of the church there was a huge font for holy water. It was shaped like a giant ciborium and was seamlessly sculpted from a chunk of limestone rock. It was always full to the brim with holy water from which Mammy would regularly fill up her pint bottle. This bottle had pride of place on a shelf by the back door of our house and we were liberally doused with holy water whenever we were setting out on a journey. Even though it was done with a great deal of hilarity, it set the tone and kept us safe.

Every second year the 'Mission' would take place. The Jesuits would come and frighten the living daylights out of all of us with heavy doses of 'hellfire and damnation'. Of all the missioners that came to Lixnaw, one preacher stood out above all others; a tall priest in black robes with beads tied around his waist. This man had fine-tuned his preaching skills. With a slow deliberate pace he climbed into the beautifully decorated pulpit of Lixnaw church, paused for what seemed like a very long time and then began to speak in a barely audible voice: "For what does it profit a man if he gains the whole world and loses his immortal soul". Through the sermon his voice would ebb and flow and then build to a powerful crescendo. The tirade lasted for most of an hour, sometimes beseeching, sometimes threatening, as he leaned out over the pulpit with beads of sweat showing on his forehead. The congregation relished it; this was 'real value for money'. For weeks afterwards Mammy talked about this sermon; loving the language and the vehemence with which it was delivered. For her a Mission without a bit of fire and brimstone wasn't worth attending.

Mammy took her religion far more seriously than the Galvin side of the family. Attending Novenas and doing the nine Fridays and five Saturdays was seen as a requirement for good Catholics. But she also loved its social side; she loved meeting people and she loved going somewhere, anywhere, it didn't matter. Mammy would take us to Confession once a month, while Daddy considered once a year sufficient. But one particular confession took the wind out of her sails and left her questioning the arbitrary nature of the penance imposed. Herself and Sara had eaten a piece of black pudding the previous day, thereby breaking the Friday abstinence from meat and hence disobeying the second commandment of the Church. Fr. Hickey was hearing confessions on this night, so Sara went in first and confessed that she had eaten black pudding on Friday. All went well; "Absolvo te... and say six Hail Marys for your penance". Sara came out and Mammy went in and confessed the same sin, that she had eaten black pudding on Friday. The man behind the screen lost the head and declared, "There's too much of this kinda thing going on in my parish, for your penance say six full rosaries". By the time Mammy was finished saying her penance she had pains in her back and pains in her brain. The moral of the story was: for a lighter sentence get into confession first.

My own first experience of confession in Lixnaw church was equally perplexing. This was my second confession; about a month after making my First Communion. On entering the Confession box, the door blocked out all light leaving me in total darkness. I didn't know where to kneel or which wall to face. I felt around the walls and finally found a wooden ledge. In my confusion I never waited for the shutter to open but rattled off my sins and came out. Kevin asked me what penance I got and I said he didn't give me any so he said I'd have to go in again for it to be a proper confession, but I'd had enough for one day. Some years later Tom also had a very memorable confession experience in Lixnaw. It was Christmas time and there was a big crowd waiting outside Fr. Molyneaux's confession box. Fr. Molyneux was getting old and was a bit deaf. When Tom went in he whispered about something he did and because the priest was hard of hearing he responded loudly, "You did what, boy"? The entire church had heard and must have wondered what dreadful thing Tom had done. When confession was over he made a hasty exit to avoid the gaze of the onlookers.

During the school term we would go to confession in Kilflynn Church. Usually there were two priests there to hear our confessions, Fr. McCarthy and Fr. O'Brien. Fr. McCarthy was always very quick and gave the regular penance of three Hail Marys. Fr. O'Brien, on the other hand, was very slow, would ask all sorts of questions and give a long penance. The church was a ten minute walk from the school so each class would be let walk there in turn. Getting there late meant going to Fr. O'Brien's box, so myself and Donal Leen would run ahead of the others and make a bee line for Fr. McCarthy's side of the church. When Donal went in to the confession box he would pull back the curtain and look out making funny faces, or sometimes we would see a hand or a leg sticking out, waving at us while we were trying to concentrate on our sins. Two of my sins were always the same; 'cursin'' and 'disobedient'. Cursing meant saying words like shit and disobedient meant not doing something straight away. I would hear Fr. McCarthy say the words 'absolvo te' as I started my devout act of contrition and before I'd be half way through it the shutter would be closed with a snap.

***

When I was ten, Mammy and Sara took me with them on a pilgrimage to the Marian Shrine at Knock in County Mayo. The Story of Knock tells of an apparition of Our Lady, St. Joseph and St. John on the church wall in the year 1879. Fifteen people from the village, whose ages ranged from mid teens to mid seventies, are believed to have witnessed this apparition. Down through the years Knock has been a place of pilgrimage for millions of people from all parts of the world. When Pope John Paul II came to Ireland in 1979 he visited Knock to commemorate the centenary of the apparition. He celebrated Mass in the new basilica that was built for the occasion. Mother Teresa of Calcutta paid a visit to the Shrine in 1993.

During the nineteen fifties the Knock pilgrimage was an annual diocesan event that took place in early summer. The old steam train ran from Tralee to Claremorris, stopping at many towns and villages along the way. A big crowd got on at Lixnaw but Mammy and Sara and four of their friends secured a compartment for themselves. This was what they wanted because it afforded an opportunity for three rosary recitals on the journey up, as was the custom. This would be repeated on the way back and, of course, three more rosaries were said walking around the shrine at Knock.

Being the only child in the compartment, I make up some excuse to escape and walk along the narrow corridor where three girls ambush me. They are Helen Flaherty, Peggy Corridon, and Mary McCarthy who have also made a getaway from their own compartment. I feel shy and awkward in their presence so I run off down the corridor but to my horror they are running after me. They see me as an escaped prisoner and they are the jailers in pursuit. When they capture me I'm manhandled and marched back to the starting point. And so our game begins. They count out a twenty second head-start and then give chase. It's exhilarating; I can barely contain my excitement as I search for hiding places along the train. But I don't run too far because I like being caught and dragged back to base, especially by Helen with her long dark hair, her bright blue eyes and her infectious laugh. I love being her prisoner. The four hour train journey goes by very quickly and our game comes to an end. I follow Mammy and Sara as they head for the Shrine and I walk around it mumbling my rosary responses, amidst the cacophony of a thousand other voices. Occasionally, I see my jailers circling the shrine with their parents and when our eyes meet a secret message passes between us and Helen smiles her bewitching smile. Her smile lingers in my mind as I walk along trying to concentrate on the rosary. Any religious fervour I might have had is gone; I can't wait for the journey back home.

As well as pilgrimages, visits to blessed wells were also part of our religious practice. There is a blessed well in Kilflynn on the northern bank of the Shannow river, whose origin dates back to monastic Ireland when a hermit monk lived there. When Fr. Hurley took up duty in the parish he revived the old tradition of visiting this well on May Eve. A huge crowd gathered on Shannow bridge and he led the procession back along the river bank. We all sang *Bring Flowers of the Rarest* and other hymns as we walked along. It was a joyful experience made more memorable by the glorious sunshine. After saying some prayers he blessed us with the water from the well and gave a short talk about *Naomh Flainn* who had built a shack nearby fourteen hundred years before. The sun was still shining brightly as we made our way back to the village. Mammy was deep in conversation with some other women. She was in her element. Meeting and talking was all she needed, but we needed more and Condon's shop was the place to get it. She gave a shilling to each of us to spend at will. We indulged our taste buds with all the cheapest junk and then walked home feeling happy and contented. It was a wonderful community occasion that brought joy and happiness to many people. When Fr Hurley left the parish the practice ceased. Some members of the clergy felt that such worship was more associated with pagan Celtic tradition than Catholic Church tradition.

After Sara's eye operations Glendaghalan Blessed Well became a regular destination for many months. The water bubbled out of the well as pure and clear as crystal glass. It was widely believed that it had curative powers for sore eyes or impaired vision. Sara patiently bathed her eyes and prayed for a miracle and we prayed with her. From there we would travel on to the blessed well at Ballyheigue, situated near the quiet, picturesque village that leads down to the beach. While the well at Glendaghalan was in the middle of an overgrown field, the well at Ballyheigue was situated in a manicured, landscaped garden with a big statue of Our Lady in the centre. Visitors walked around the circumference, enjoying the beauties of Nature and whispering prayers in private. Invariably, Mammy would meet some of her many cousins. Her mother's family, the Hanlons, had produced a squad of first-cousins that all lived in the locality and nothing delighted her more than an accidental meeting. They would walk along the beach or relax on the promenade and talk for hours on end. Mammy loved the sea. She had grown up near the little fishing cove of Meenogahane about five miles north of Ballyheigue. The sound of the rolling waves was in her head and kept calling her back. So visiting the Leen's family home was a regular trip for her and for us; a wonderful and exciting experience that we enjoyed and cherished.

Chapter Thirteen

The Leens

The Leens were the other side of our family. They lived beside a beautiful fishing cove called Meenogahane; a transliteration from the Gaelic name of *Mín Uí Catháin,* meaning Keane's cove. Who this Keane person was nobody knows; the history has been lost and no family bears that name in the locality. Instead the name Leen is the dominant family name in the area. Mammy was the eldest in the family of six sons and two daughters born to Johnny Martin Leen and Hannah Hanlon. Her name was Margaret, but everybody called her Madge, and her brothers were Martin (aka Chris), Jack, Tom, Danny, Mikie and Pat, and her sister, the youngest in the family, was named Annie. The Galvins and the Leens were well matched and got along well together. The two different family backgrounds complemented each other and brought a richness of experience into our young lives. Our two youngest uncles, Pat and Mikie, then in their early twenties, would visit Crotta regularly, making the long trip on an old-fashioned bicycle. We loved to see one or other of them coming through the road gate because they always spent more time playing with us than talking to the grown-ups. Their older brother Tom had inherited the family farm and the middle brother, Danny, was then a missionary priest in the Philippines. The eldest brother, Chris, had emigrated to America at a young age, enlisted in the army and was killed in action. The second eldest, Jack, had contracted meningitis in his late teen and was left severely incapacitated. He needed constant care and attention. It fell upon Madge to be his carer until he died in his early twenties.

In the summer of 1944 Madge was visiting her cousin, Mary Whelan, nee Leen, who lived in Ballyrehan near the Galvin homestead. She was in her prime and unattached. Mary Whelan's husband, Gerry, seized the opportunity to try his hand at matchmaking and introduced his bachelor friend and neighbour, Jeremiah Galvin, to his visitor. She was in her mid twenties and he was forty four. Despite the age difference, the relationship flourished and they were married within a year. They soldiered through life together, taking the ups and downs of life with stoic bravery and an unfailing trust in God. They both lived well into their nineties and enjoyed good health to the end.

It was difficult to gauge their love. Like many married couples of their generation they never expressed their feelings openly in words or never showed any outward signs of affection in public. But some spark glowed right to the end. When she knew he was slipping away, she whispered "Don't leave me" and in a muttered voice he replied, "Sure, didn't I stay with you a long time". A few days later he slipped into a coma and we all stood around the bed as he breathed his last. It was the only time I was present at a death-bed. I watched his open mouth gulp in the air at ever-increasing intervals. As the pause between each breath grew longer I kept thinking each one was the last, and then they stopped and he was gone. Daddy was dead. As the mystery of death filled the room I tried to take solace in Christian belief that he was gone to a place or state of happiness. I tried desperately to hold back my tears but my throat was choking as memories started to flood back. Mammy kissed him and said goodbye, as we all stood there saying our private farewells to a good father. She outlived him, as might be expected, by twenty years. She died at the age of ninety-six and was laid to rest beside him in Kiltomey cemetery in July 2011.

Of all our uncles, Pat was the favourite. He was the second youngest in his family, a year and a bit older than Annie. In his early twenties he realised there was no future for him in Meenogahane so he decided to take the emigrant ship. He worked for many years in London before moving to Chicago, where he met Peggy and raised their three children. During his time in London he would come home every summer and jazz up our lives for a few weeks. He had a great sense of humour and loved telling jokes and funny stories and messing around with us as children. He was more like a big brother than an uncle and when the horse-play got out of hand in the kitchen Mammy would come at us with the wooden spoon, but Pat would stand between us and retribution. I can still hear him say; "Arrah! Madge, leave them be, what harm are they doing". She would only pretend to be angry because she had always been more of a mother to Pat than a big sister.

He liked spending time in Crotta and got on well with Daddy and Mick. He would share a double bed with Mick in the downstairs bedroom. The following day he would accuse Mick of lying on his shirt-tail all night so that he couldn't turn over. This sort of sparring between them would go on through the whole day. Pyjamas hadn't yet found their way into our house; the shirt you worked in was the shirt you slept in and nobody gave it a second thought. If Pat slept late any morning Sara would enter the room and attempt to pull him out of bed. His screaming could be heard for miles as he tried to hold on to whatever bit of dignity was left. All in good fun, Sara loved him to bits; there was nothing she wouldn't do for him, or he for her. He held Daddy in very high regard and they had a very close relationship. They would travel to Croke Park together if Kerry made it to the All-Ireland final and, on returning, relate their adventures to our amazed ears. But most of his time at Crotta was spent working. Pat was no stranger to hard work and his help was greatly appreciated.

It's the summer of fifty-five and our turf supply has run out. Pat volunteers to replenish the stock with a horse-rail of new turf from the bog. So the faithful Moll is harnessed and stands patiently between the shafts. Pat asks if we would like to go with him. Mammy is a bit worried but Pat reassures her that we'll be safe, but Tom is too young and has to stay home; he is left behind with tears in his eyes. As Moll trots down the Lixnaw road Kevin and I feel very grown up standing beside Pat in the big common car peering out through the rails. When we get as far as Donal Lynch's shop in the village Pat reins in the horse and, without a word, climbs down out of the rail. He spends a long time in the shop and when he comes out he is carrying a brown cardboard box and we wonder what's in it. But without a word we continue on our journey. He has difficulty finding our bank of turf because all the banks of turf are identical. But then he spots the bog-cart that was used for getting turf out to the road. The bog-cart is a homemade big box with very broad wheels. These prevent the load of turf from sinking. The bog is too soft in some places for Moll to walk, so Pat has to guide her carefully as she pulls the bog cart out onto the road. Kevin and I have great difficulty throwing the sods into the horse-rail because the sods are big and the rail is high but we are determined to prove our worth. This routine is repeated many times; filling the little bog-cart and unloading it.

While we are working two skylarks are serenading each other high up in a cloudless sky. Their harmonious melody is filling the air. Not another sound can be heard on this vast open wilderness. It is difficult to see them clearly in the dazzling light of the noonday sun, only a flutter of wings is noticeable but their rapturous song is coming at us from all directions. Along the bank, the bog cotton flutters in a warm breeze that wafts the perfumed scent of heather in our direction. The scene is perfect; too good to be missed, so Pat ties Moll to the horse-rail and brings the brown cardboard box to a flat, dry patch of ground. An eye-popping surprise awaits us; there's lemonade, cream buns, different kinds of biscuits, bananas, chocolate and sweets. We feast like ravenous wolves while gulping down the fizzy lemonade. Pat tells us to slow down, so we enjoy our mouth-watering picnic in the unspoilt beauty of the open bog. In a little while he lies back against a tuft of tall grass and smokes a Sweet Afton cigarette. The smell of the tobacco adds to the sensual delight of the occasion and I ask him for a puff. It makes me cough and brings tears to my eyes and makes me realise that tobacco smells better than it tastes. We sit there for a long time in our peaceful surroundings. Eventually we get the horse-rail full and we head for home. When we tell Tom about our exciting day he turns to Mammy and says in a sad voice, "What a pity I couldn't go to the bog" and the phrase and the story lived on in our family folklore for years to come.

That summer Pat plastered the rough stone wall of the stalls where I had split my forehead as a small child. Plastering was his trade in London and he was good at it; with hand-hawk in one hand, plastering trowel in the other and a Sweet Afton between his lips he could slap on the mortar at great speed. It was fascinating to watch him work as he tossed the mortar from trowel to hand-hawk and back again. When that job was finished he took on the difficult task of plastering the walls of the back bedroom where the three-hundred year old lime-mortar has begun to crumble.

This job takes longer than anticipated. He finishes floating down the wall by candle light and I'm the candle holder, following his movements along the wall with my hand raised high. Pat keeps asking if I'm okay and I reply "I'm okay, Pat", but I'm not. I'm very cold and my arm is aching, but a good marine doesn't desert his post, not when it's for Pat. But time drags slowly and I'm wishing he wasn't such a perfectionist as he meticulously floats down the dimly lit wall. Some days later I come down with an illness and have to stay in bed. Pat comes to see me to say goodbye. He's all dressed up and ready to go back to London. He sits at the end of my bed and talks about everything except London. I know he hates going back but this is the life he has chosen. As he says goodbye I feel my eyes stinging and I turn my head away. He tries to make some funny remark and then unties the little silver watch chain that hangs from his waistcoat pocket. He insists I keep it to remember him. Many years have rolled by; Pat is dead and gone, but his watch chain, now tarnished and grey, is still a treasured possession, amidst life's other bits and pieces, in the bottom drawer of my desk.

***

Uncle Danny came home from the Philippines in the spring of '58 after a ten year stint as a missionary priest. Mammy was beside herself with excitement waiting for him to visit. They had been very close growing up and now he was perceived in a heroic light. Having a brother a priest was an honour for the family and had raised the Leen family status throughout the community and further afield, especially amongst the clergy. And there he was in our back yard; a striking figure, tall and athletic-looking in his black suit and white Roman collar. Mammy had told us we should address him formally as we did with the priests of the parish; "Yes Father/ No Father", but he laughed and told us to call him Danny. However we were a little bit in awe of him and it took some time before we could relax in his presence. But an exciting summer awaited us. His brother Tom bought him a second-hand car, a black Ford Anglia, and we had numerous trips in it to football matches, days by the sea at Ballyheigue and Ballybunion and shopping trips to Tralee and Listowel.

And of course we spent a lot of time with him at the old homestead in Meenogahane. These occasions were great family gatherings for the Leen family. The kitchen could barely fit the crowd; there was Mammy and Annie and three brothers, Tom, Danny, Mikie, four young Galvins and Helen. Helen was Annie's eldest daughter who lived in Meenogahane at that time. I was the same age as Helen and we were good friends. As children we loved listening to the adults talking, but on this occasion Helen and I grew tired of the conversation and went over to the big yard by the hayshed to play with my new tennis ball. This tennis ball had come last Christmas from Lil Shanahan in London but I had taken such good care of it that it still looked brand new. As Helen and I were playing a 'game of donkey' with the ball, Danny, Tom and Mikie came strolling along; they were going for a walk back the fields. Danny had been a talented footballer during his college years and as I threw the ball to Helen he leapt into the air and caught it and with exaggerated poise and showing off his skill gave it a mighty kick over the hayshed. He chuckled at his achievement and walked away with the others while Helen and I went around the hayshed to search for my precious tennis ball. We searched everywhere but could not find it. I felt peeved and wished that Danny could have shown off his football skill in some other way.

It's Sunday evening and Helen hasn't finished her homework for school the following day. She has to write a composition about 'The Watchmaker'. She searches in her schoolbag and pulls out a dog-eared copybook. Her pencil needs sharpening so Mikie takes out his pen-knife and pares the pencil into the open fire. She had written one sentence earlier: "The watchmaker lives in Tralee" and then the Muse had deserted her and she put the copybook back in her schoolbag. Now a deadline has to be met. She sits there chewing the top of her pencil but no inspiration is coming. She looks at me with pleading desperation in her eyes. I'm thinking how much I hate doing my own homework, so why should I take on more. The adults are deep in conversation and oblivious of Helen's need for sentences about a watchmaker. I try to help but the topic is so removed from anything I know that it takes a long time. Together, we struggle to compose nine or ten stilted sentences and she tries to spread out her writing to fill up the page. With a sigh of relief she stuffs the copybook back in the bag.

After supper the adults stay talking in the kitchen but Annie asks us to accompany her to the bonfire. Annie seems happier in the children's company than in the adults' company. It's May Eve and the lighting of bonfires is an old custom to see out the winter and welcome in the summer. Some people have started the fire earlier a little way up the road so Annie organises us to bring materials to burn; big sods of turf and some old boards. We build up the bonfire into a high conical mound and wait. It doesn't take long. The dry boards begin to flame and with loud crackling explosions send sparks flying in all directions. As the heat grows stronger we have to move back a pace and our faces are taking on a red sheen.

Annie tells us about the fairies being on the move on this night and to watch out for magic happenings. She has brought a long iron rod for poking the fire and at one stage she leaves the end of it sitting in the intense heat and we watch its colour change from deep red to bright red. When it is glowing she carefully pulls it out and strikes it against the rough stone wall nearby, setting off a shower of sparks against the darkening sky. We gaze in awe at the magic display. One end of the rod remains cold so she allows us take a turn at creating the fireworks. We reheat it again and again. By the end of the evening the straight iron rod has become twisted and tortured and shaped into a hook at the top. It has taken on a human form. When it cools down Annie adds two arms made of rusty barbed wire and tells us that iron man will protect the village from the fairies for the rest of the year. Later that night, as Mammy drives us home the bonfire is still glowing faintly. Everybody has gone but the iron man is standing by the wall where Annie had secured him. Like a faithful sentry, he will stand guard at his post until next May Eve.

Down by the seashore in Meenogahane was a magical place for us; picking periwinkles, gathering dilisk and jabbing the *bárrnach* (limpets) off the wet rocks. Mikie was the one who always took us down to the pier. He loved showing us what to do and where to go. There was nothing about that area of sea and rocks that he didn't know. He knew the best rock-pools for swimming and the best places for fishing. He would make up little fishing lines for us as children with which to catch 'heart-heads' under the big rocks when the tide had gone out. Back at the house he would rake out some hot cinders from the open fire, on which to cook the *bárrnach*. We carefully placed each shell on the hot coals. They were ready in five minutes and were so delicious that I often burned my tongue, licking the inside of the shell for the last bit of salty flavour. He would give us a bag of periwinkles to take home to Sara and Lizzie. The following day, those two good ladies would sit by the kitchen table, armed with a straightened-out safety-pin, until every last periwinkle had been eaten.

Danny visited Crotta regularly and spent endless hours talking to Mammy. When we saw the black Ford Prefect pulling in to our yard other distractions were put aside and we made a bee line for the house. We liked to observe him and listen to his conversation. He smoked tipped cigarettes and looked the cool clean hero. He had brought a huge quantity of cartons from the duty-free shop at the airport. I loved the aroma of cigarette smoke as he nonchalantly exhaled a bluish cloud around our kitchen. Resistance was futile; I had to try it. At first I found a few stamped-out cigarette butts and sneaked off with a spare box of matches to the back stall. After a lot of puffing and coughing I got the hang of it and swaggered around with the butt in my mouth. On another occasion Danny lit the tipped end of a cigarette and had to throw it away but we noted the spot and retrieved it later. Kevin, Tom and myself shared it, as we walked through the fields, feeling very grown up.

As the summer drifted by, Danny would disappear for weeks at a time and things would get back to normal, but on one occasion he returned unexpectedly accompanied by two women, one a cousin from Ballyheigue and the other a very glamorous young woman, whom he referred to as the Kiwi Girl. Even as a prepubescent youth I was keenly aware of her beauty. She was tall and elegant with bright sparkling eyes, an exotic appearance and a polished accent. Mammy prepared tea in the parlour and Danny asked me to play a few records on the gramophone. He had a favourite John McCormack record called: *When You and I Were Young, Maggie,* which I had to play over and over again. When they had finished eating, he asked me to play the 'dance' record and he entertained everybody by throwing off his jacket, undoing his clerical collar and dancing a hornpipe. Then it was time to go. I was 'hanging in there' as they sat into the car in the hope of getting a spin into town or somewhere exciting, but no such luck; they planned to tour around South Kerry and West Cork for a few weeks. Danny continued with this 'playboy' lifestyle until one evening in September he announced that his leave-of-absence was over and that he would be leaving the following day. Mammy cried and hugged him again and again and we stood around not knowing what to do or what to say. He would not be going back to the Philippines; instead he was going to America. After one final hug he sat in, started up the engine and drove slowly down the yard. We stood there watching and waving. He bipped the horn twice as he rounded the corner; it was his last farewell. Danny was gone and would be gone out of our lives for a long, long time.

At first he went to San Francisco where his brother Pat was living at that time, but after Pat had moved to Chicago Danny's letters stopped coming. He had disappeared off the radar, gone without a trace, leaving everybody bewildered and worried. Years went by and even though Pat had hired a private detective to shed some light on his mysterious disappearance nothing came of it. Mammy feared the worst and would pray for him every night at the rosary. She believed that he was dead but to everybody's great surprise he re-surfaced many years later with his wife and children. When word filtered back home some members of his family were angered and outraged and dreaded the humiliation of facing the neighbours. All the neighbours had rejoiced and celebrated his return from the Phillipines in '58 and the family had basked in the glory of having a priest in the family. Catholic dogma stated that a priest was a priest for life and now Danny was married and had taken up a ministry with the Anglican church in America. This was a bridge too far for some members of his family and so the book was closed; the shameful outcome would have to be kept secret.

Whenever neighbours asked about him they only got some vague response; that he didn't keep in touch and contact was lost. But *as the hare whom hounds and horns pursue, pants to the place from whence at first she flew*; Danny was drawn back to see the old haunts. One summer a story had gone around about a visitor wearing dark glasses and a pulled-down hat and some neighbours speculated that the 'yank' standing on Meenogahane pier, looking out to sea, looked very like Danny Leen. The family denied all knowledge and didn't want to hear about it nor engage in these conversations, but the passage of time changes perspective and shines new light into dark corners. A decade later Danny and his family visited Ireland and were warmly welcomed in our house. Mammy was overjoyed to see him and became very good friends with his wife Barbara and we were all delighted to meet our new yankee first cousins, Virginia and Kevin, but the old homestead at Meenogahane remained out of bounds for them.

***

The autumn of fifty-nine was "made glorious summer" by the endless weeks of sunny weather. The usual work went on and I was now trying to fill the lead role as helper to Daddy and Mick and finding it very difficult. Kevin was gone; packed off to boarding school in St. Brendan's College. But this was not Kevin's first time away from home. In fact he had been gone all summer. In order to acclimatise him to being away from home Mammy thought it would be a good idea for him to spend some time in Meenogahane with his uncles, Tom and Mikie. The two of them were living a bachelor existence in the family home in which Mikie had assumed responsibility for running the house while Tom organised the farm work. So Kevin pioneered the Meenogahane exile experience, which I would try out the following year. When he returned, he told such wonderful stories, about fishing and going out in the currach to check the lobster pots, that for a whole year I was looking forward to spending my summer there.

So I was taken to Meenogahane around mid-July of the following year for a six week stint away from home. A cluster of six families made up the little village and all must have had common ancestry for their names were either Leen or O'Connor. The houses were in very close proximity to each other and the farm buildings were a complicated, higgledy-piggledy mixture of sheds, strewn here and there. Farming was their way of life and their small farms went to the edge of the sheer cliff that looked out over the Atlantic. There were no trees to be seen anywhere except a few hardy hawthorns that survived in a sheltered glen by keeping their heads down. These trees provided a useful canopy for animals from the summer sun or winter snow and from a distance looked like a group of men with severe back pain. Because the dwelling houses were built so close together neighbours kept meeting each other constantly during the course of the day. Each time it was difficult to think of a new greeting, but luckily the weather was very changeable and provided a different comment every time they met.

In one of the houses, a girl named Noreen had also come to stay with her uncle for the summer holidays. She seemed to be good friends with Mikie by the time I arrived and liked to accompany him as he worked around the yard. She had shoulder-length brown hair, bright blue eyes and a pretty freckled face. I had just turned thirteen and she was twelve and a half. Hanging out with girls was not my thing at that stage but circumstances had thrown us together. She was tomboyish enough for joining in whatever work was going on so we gradually came to accept each other. She would often accompany Mikie and myself down to the pier for periwinkles or back to the hill field when he was checking on cattle. One day after checking the cattle we walked along the cliffs as far as Beenatee. The cliff face at Beenatee, is a sheer drop of hundreds of feet and Mikie showed us where he and Tom had daringly lowered themselves down on ropes to salvage shipwrecked cargo many years before. It was a scary thought, made scarier by the thunderous roar of the waves that reverberated along the cliff wall as we walked along.

Noreen is dressed in denim jeans with straps over her shoulders like workmen's overalls. Her auburn hair is tied back in a ponytail that swings hypnotically as she walks. She is full of energy and keeps messing about with Mikie, so together we conjure up a game of how to topple him to the ground. She leaps on his back as I try to drag him down, but he is strong with a good centre of gravity and we fail to topple him, so he runs ahead of us. As we chase after him, he gives way to youthful abandon, disturbing the seagulls as he mimics their shrill screams. Running and jumping and messing with Mikie we journey on. He has a good knowledge of the different seabirds, the various types of gulls and terns, and shows us where they nest along the cliff face. We gaze in wonder at their grace and beauty as they float in the air and then swoop downwards to land on the water. The edge of the cliff is covered with beautiful sea-pink but he warns us not to go too near in case the edge might give way beneath us.

We come to a sheltered spot and lie against a soft grassy ditch and take in deep breaths of the salty air. The vast Atlantic is stretching far and wide beyond the curve of the dark cliff. The seabirds are nesting precariously all along its treacherous ledges. The blue water sparkles in the evening sun and a gentle breeze stirs the tall grass at our feet. I close my eyes to relish the beauty of the moment but Noreen's presence is distracting my thoughts. She has thrown herself against the grassy bank only inches from where I'm sitting. I'm conscious of her breathing, as she exhales through puffed lips, dispelling the stillness all around us. I'm afraid to move for fear of brushing against her. I remain motionless for what seems like a long time. But then the spell is broken and the moment passes. Mikie sits up and lights a 'John Player' cigarette from his packet of ten. I take the box from his hand to study the sailor's face printed on it and ask if I could try one. He says I'm too young but hands me his cigarette for one puff. I splutter and cough and Noreen laughs her beguiling laugh.

But my summer in Meenogahane wasn't all fun and games. There was a lot of hard work very similar to what I had left behind in Crotta. The day started with the milking of the cows and then I would travel with Mikie to the creamery in the donkey and cart. After a few days, when I had learned the routine, he let me off on my own to the creamery but the donkey took most of the responsibility. He had his own set pattern. He would run bits of the way; would overtake other slow-going creamery-goers and knew where to join the long queue when we got there. I was only a *garsún*, so willing hands would help empty the milk churns into the big stainless steel vat for weighing. Some of these men were friends of Mikie's and would ask about him. They all seemed so good humoured as they cajoled and joked with each other as the queue moved slowly forward.

On the way home the donkey knew where to stop. He would turn in to Lawlor's house by the forge where Old Maurice Lawlor would always greet me with the same words: "How's the hardy hooker?" I can only assume that it was a seafaring expression from his younger days and didn't have any other connotations. He would hand me a galvanised jug for skim milk, which I would fill from the churn. Back at the house I would reverse the cart into the shed and unhitch the donkey, removing the collar, straddle and britchen. The winkers would remain on till we got to his field. If Noreen was waiting she would leap-frog onto his back behind me and together we would gallop him all the way back the road. Staying on board was difficult and amid her excited yelps she would tear at my shirt for support whenever she slipped sideways. We were both amazed at how fast this poor old donkey could run. I guess he knew a whole day of freedom awaited him with plenty of grass to eat. His daily job was done and he would laze around in his field till the following morning.

Sometimes Noreen doesn't show up and it casts a gloom on the whole day's work. She's been missing now for two days and we think she may be gone home. It's a showery afternoon and Mikie and I are cutting ragworth weeds in the donkey's field with slash-hooks. Ragworth is a noxious weed that multiplies rapidly if not kept in check. The field is a sea of yellow, the ragworth is in full bloom but the seeds have not yet ripened, so this is a good time to cut them. We are working away steadily when she appears at the gate. My heart skips a beat as I see her running down the field. She is wearing her denim overalls and does a few cartwheels as she approaches us. She greets us with her charming smile and asks if she can help. There's no spare slash-hook for her so Mikie suggests that the two of us should pull the smaller weeds by hand and let him deal with the big ones. The pulled ragworth brings a weighted root-ball with it and this can be hurtled into the air with a swing of the arm. So this becomes our game. As we pull each ragworth we count one, two, three and let fly into the air to see who can throw it highest. Noreen lets out a yell each time she launches her missile. She quickly develops a skilful swing and is outdoing my best efforts. The breeze blowing in from the sea keeps tossing her hair across her face and without stopping she throws her head back to clear it away. Her swinging technique is poetry in motion and she's so totally engrossed in the game she's unaware of my gaze. I've lost interest in the contest; I don't care who's winning any more, I just want to watch. I'm enthralled by the speed and grace of her movements and the joy of her excited screams. Then suddenly the rain comes pouring down to break the spell and we dash for shelter.

Heavy showers are falling at regular intervals so we have to run for shelter to the donkey's shed. This makeshift house is built with rough planks and rusty galvanised sheeting. Mikie lights up a cigarette, smokes half of it and saves the other half for the next shower. The swallows keep swooping and diving outside the door trying to gain entry. They're not happy with our presence in the shed because they have hungry chicks in their little mud nests and we are preventing them from flying in and out. Mikie explains how they make the nests by mixing mud and spittle in their mouths and bit by bit shaping their little mud home. He tells us it becomes as hard as cement and he climbs up on the donkey's manger to feel one of the nests. As he does this, out flies a swallow into his hand, nearly knocking him off the manger. The poor little creature is frightened and pecks at his thumb but Mikie holds it carefully and marvels at the fact that this little tiny creature has flown all the way from North Africa. He tells us that the same swallows come back every year so Noreen finds a piece of red thread in her pocket and together they secure it around the swallow's leg. "See you next year, little swallow", she whispers and wonders if her piece of red thread will make it all the way to Africa and back again to the donkey's shed in Meenogahane. Mikie opens his hand and the swallow flies away.

Tom Leen was a very progressive farmer who got maximum return from his small farm. The horses had been sold and replaced by a Fordson Dexta tractor and trailer. Turf was also their source of fuel for the winter so one day we headed off early for the bog to draw out the turf to the roadside with the new tractor and trailer. A neighbour, Bill Leen, came with us but

Noreen was not allowed come. That morning I had pulled on my wellingtons without wearing any socks and for this omission I was to pay a high price. The bog was dry enough for the tractor and trailer to travel on the turf bank, so Tom drove the tractor up the long bank while we walked behind. We did this all day long, filling and emptying the loads of turf. The day had become very hot and my feet were sweaty and sore in the wellingtons but a macho sense of pride prevented me from saying anything, especially as Bill Leen kept teasing me to "Get a move on". By evening I was in distress and in serious pain so Mikie told me to rest for a while. When we got home he brought me a basin of hot water to wash my feet and he got a shock when he saw the state of my skinned and bloodied toes. Like Mary Magdalen he tenderly bathed my feet and massaged them with soothing oil. He then rubbed in some antiseptic cream and covered the worst parts with bandages. I sat by the fire and read my comics and got some pleasure from the thought that I would not be expected to do any work for a few days.

At eight o'clock, as was his habit, Bill Connor came in the door with his usual "God bless all here". He was a very tall thin man, in dark clothes and wore exceptionally big shoes. His accustomed spot was waiting for him at the opposite side of the open fire. He would lean back his chair until his head touched the wall and balance himself, precariously, with those enormous big shoes touching the floor. Tom and himself talked in low measured tones about the bog and other matters of interest to farmers but in between there were long silences with only the hypnotic ticking of the clock over the fireplace. I would read my comics and Mikie could often be seen darning his socks. Around ten o'clock, Bill would say, "I think it's bunking time", and leave. It was bedtime for Tom and Mikie too. They didn't say the rosary like we did at home but knelt by their chairs and said their own individual prayers, so I followed suit and rattled off some *Hail Marys* to myself. Mikie would stay behind and tidy things up around the house. It was he who always got the meals ready and washed up afterwards. He was a good cook; I enjoyed his meals; lamb chops instead of boiled bacon and sometimes fresh mackerel fried in a flour batter and some days he would make delicious soup with fresh vegetables. Noreen would sometimes pick carrots, peas, parsnips and turnips in his well-tended garden as I was releasing the donkey into the field nearby. The fresh young peas were so sweet that by the time we got home to Mikie we had most of them eaten.

The following week after the bog finds us drawing in the hay. As the hayshed fills up it is difficult to squeeze in the last of the hay. There are three stages to getting the hay up to the top of the shed. Tom, on the ground, pikes it up to a ledge where Mikie is standing and Mikie pikes it up to a higher ledge where Noreen and I collect it by hand in big bundles and drag it back to the end of the shed. She uses my catch-phrase, "make way for Lord Kitchener", as she squeezes by along the narrow passage with her bundle of hay. At first it's great fun but the day is hot and the corrugated iron roof over our heads is turning the top of the shed into an oven. The space is filling up fast and we are stumbling and falling as we pack the hay into every corner. It's exhausting work under terrible conditions but she's making me laugh every time she falls and it eases the pain. In spite of sweat, heat and the dust we struggle on, but finally we shout at them saying there's no more room, so they stop throwing it up. Mikie begins to fill up his own ledge and we take a well-earned rest. We both need fresh air so we crawl to the side edge and poke our heads out through the iron bars that support the roof. We lie there side by side for a long time, like prisoners in a medieval stock, our disembodied heads sticking out through the bars and causing great amusement to those down below. We are still hot but a salty breeze blowing up from the sea is beginning to cool us down. Noreen wipes the sweat from her forehead, she protrudes her lower lip and blows a breath of air upwards to cool her flushed face and smiles her bewitching smile.

On sunny evenings when work is done Mikie and Patty Connor go down to the pier. They launch the black, bitumen-coated currach into the water and row out a few hundred yards to check their lobster pots. This is fiercely exciting for me but also a bit nerve-wracking as the waves outside the harbour wall are quite strong and the currach bobs about like a cork in the water. Both of them do the rowing while my knuckles grow white as I tighten my grip on the cross-board I'm sitting on. The lobsters are valuable and not for eating but the giant crabs are of no commercial value and are eaten when we get home. We visit all the lobster pots, each clearly marked with a yellow buoy, and some evenings we're in luck but most evenings there's nothing in them except the crabs. When they finish checking the lobster pots, they throw a fishing line with many hooks into the water. A great deal of patience is required but the prize is worth waiting for; a few fresh mackerel for supper. Mikie is a dab hand with the frying pan and nothing in the world tastes as good as fresh mackerel fried very crispy in butter and salt.

In late August Eamonn Diggin pulled his threshing machine into Bill Connor's yard where two big ricks of corn were standing. When the threshing machine arrived it became a community enterprise just like

Crotta; the *meitheal* gathered and followed the threshing machine to each farmyard until the job was done for all. Bill Connor had a big house with few occupants so he stored the grain in one of the downstairs rooms. The sacks of grain were being stacked in two layers by Stevie Connor and Tang Sullivan. Noreen and I were helping them. We were up on the first layer of sacks, settling the second layer into place. The sacks were heavy but we had become a good team and worked well together. While waiting for more sacks to be brought in the two men amused themselves saying that I should kiss Noreen. They grabbed us, one each, and dragged us down off the bags of grain. We kicked and struggled but they were far too strong. They pressed the two of us together till our faces met and held us there for what seemed a long time. It could not have been classified as a kiss but the close body contact left both of us feeling awkward and embarrassed. The men continued teasing and laughing and thought it was very funny so we laughed too, but our daily encounters were never quite the same after that. We never spoke of it but something inexplicable had come between us that we didn't understand.

As the summer drew to a close, we only had a few days left and I wanted so much for her to be there. Each morning I waited in hope but she didn't show up and those last few days were dull and empty. Then Mikie heard from her uncle that she was gone and my heart began to sink. For the rest of that day I felt a dull ache in my chest that wouldn't go away. My Noreen had vanished like a wisp of smoke. Meenogahane had lost its glow and I just wanted to go home. Our paths never crossed again but the memories of that summer lived on and her impish grin and her carefree laugh got etched in my mind. I often wondered what became of her and how her life had turned out and on a few occasions contemplated attempts at contacting her, but lacked the courage. In my own case, my life was about to change drastically. Within a week I was being driven along the Killarney road with Kevin to St. Brendan's College, where I would spend the next five years living the monastic life of a boarder.

Chapter Fourteen

St. Brendan's

St. Brendan's was built in 1860 as the diocesan seminary to cultivate vocations to the priesthood for the diocese of Kerry. By 1960 it had changed its name from St. Brendan's Seminary to St. Brendan's College but was still known colloquially as 'The Sem'. Set in tranquil, lush surroundings, it provided segregated education for the local students of Killarney and for three hundred or so boarders from all parts of Kerry and west Cork. The ethos of the college, as might be expected, was decidedly Catholic and the teaching staff consisted of fourteen diocesan priests and two laymen. The curriculum included English, Irish, Mathematics, History, Geography, Science, Latin, Greek and Christian Doctrine. The organised structure of each day was very full and left very little free time. Classes ran till four in the afternoon except on Thursdays and Saturdays when they finished at one o'clock. We did four hours of supervised study in the big study hall every night and three hours study on Sunday mornings. I was fascinated by this new educational environment and thrilled to be part of it. Life quickly took on a very fixed pattern and I was enjoying the classes and all the new things I was learning. Being a boarder held out the promise of exciting times ahead and interesting new experiences. By the end of the first month I had made many new friends and was coping well with all subjects, but the cane, as a weapon of enforcement for learning, was an ever present threat.

The cane was used liberally in the classroom by some teachers which resulted in students giving extra study time to those subjects. Very quickly we got to know each teacher's form and worked accordingly, driven by the impulse to avoid pain. When punishment was received, the sting of the cane on the palm of the hand could be eased by clasping the cold metal of the desk or by sitting on the injured hand, but holding back the tears was often difficult. Serious breaches of discipline in the study hall were punished severely in full view of other students; sending out a clear message to all. Strict silence was observed during study time. Asking for help from a fellow student was risky as it could be misconstrued as time-wasting and could result in three parallel stripes on each palm. Most of the priests carried the cane in the folds of their long black soutanes and some had perfected the art of caning to a fine degree. Six was the magic number. Some would give all six on the proffered hand while others believed in a more even distribution of pain. Corporal punishment as a means of enforcing high academic achievement seemed to work. St. Brendan's was renowned throughout the country for its high achievers.

Outside of the classroom and study hall there was no supervision and so bullying behaviour went unnoticed and unpunished. Weak and vulnerable students were legitimate targets for the alpha males, of whom there were many. As with any large group of uncensored adolescents, name-calling and jeering were par for the course. First-year students were called 'plebs' and during the first week of college we were dragged into the washroom and 'baptised' under the cold water tap in the wash basins. This was an accepted practice of long-standing and was regarded as good fun by all except the victim. It was best not to resist because some boys were very rough and didn't care about banging an unprotected head against the enamel basin. This was the nick-naming ritual; every boy got a nickname, many of which were sadistically cruel, denoting some character trait or bodily defect.

A more playful initiation was the 'hairy belly'. There was a large expanse of lawn at the rear entrance to the college that Fr. Moynihan, the president, kept in immaculate condition. He seemed to spend most of his days on the motor mower going round and round, leaving a trail of sweet smelling grass that reminded me of the meadows at home. In the evenings as we walked the tarmac path around the lawn, with the newly-cut grass drying in the sun, somebody got targeted for a 'hairy belly'. This meant stuffing bundles of grass inside the victim's shirt and jumper until he looked like Michelin man. But most of the time we just walked round and round in phalanx formation. It was the custom to stay grouped in our own columns i.e. with boys from the same part of the county. Occasionally, we would see a senior student walking with a young junior student and somebody would shout "combo, combo", the meaning of which was lost on me at the time. This was my first impression of boarding school and even though I missed home I was generally happy with my lot.

Ten o'clock was bedtime. I slept in a dormitory called the Tower Hall which was part of the bishop's palace, a magnificent cut-stone building set in beautiful surroundings. There were twenty four beds, twelve on either side, with a locker beside each bed. A senior boy named Roddy O'Donnell was the prefect. He was tall, well built and a little overweight and every night he would launch himself onto his bed as if he were doing the high jump, until one night the wire springs snapped and he crashed to the floor. Our shrieks of laughter brought Fr. Horgan, who supervised the dormitory, rushing down the stairs with cane in hand. Like rabbits dashing for cover we quickly hid under our blankets. The lights were out so he stood at the door shining a torch around the room. Roddy lay motionless on the floor till he retreated and he slept on the floor that night.

Lights-out was at ten thirty sharp with no more talking. Roddy didn't enforce this rule and was often the greatest offender. Fr. Horgan would randomly appear at the door to check that we were all in our beds, so it was the duty of those nearest the door to keep a listening ear for his footsteps on the stairs. The word 'nixxx', uttered in a hushed tone, was the red alert and we would dash into bed and cover our heads. But one night I had been given a 'french bed'. This was a practice of removing the top sheet and doubling up the lower sheet to make it look as if the two sheets were in place. This only allowed three feet of leg-room under the covers. I was chatting with a friend when I heard the prolonged 'nixxx'. I dashed into bed but couldn't stretch down my legs. Fr. Horgan walked around the room with his flashlight. A lot of snoring and sleepy noises had suddenly begun to fill the room. I lay there motionless, with my knees touching my chin, until the inspection was over.

Every morning we were awakened at 7.15 by the loud clanging of the college bell. At 7.30 Roddy marched us up the avenue towards the oratory to be in time for Mass at 7.40. Then we went down the long corridor to the refectory for our breakfast of bread and butter. After breakfast we collected our textbooks for that day's classes and did a quick review before heading for the classroom. On Thursdays and Saturdays we had a half day, which meant we could play football in the sports field and on Sunday afternoons, when occasion arose, we were allowed to go to Fitzgerald Stadium to watch inter-club football games. We would sit on the front concrete benches, close to the action on the field of play. When the whistle was blown at half time we would drift onto the pitch to get a close-up look at some of Kerry's outstanding senior players, like Mick O'Connell, Mick O'Dwyer, Tom Long and Seamus Murphy and many others. These were our heroes and we tried to emulate their style of football in our own games.

My imaginary football skills greatly exceeded the reality of my performance, but added to my enjoyment of the game. I now preferred football to hurling and Thursday afternoon was my favourite time of the week. Sara had presented me with my first pair of football boots before coming to the Sem. They had white soles and eight white cogs for gripping the ground. I loved those boots. One particular Thursday we were playing an important match. Fr. Tom Pearce was our trainer and we knew that most of his class on Friday would be given over to analysing the game. I wanted to be at my best but on entering the changing shed I was shocked and dismayed to see that my sports locker had been raided.

Somebody has smashed the lock and taken my boots. I'm devastated at the loss of the boots and to add to my confusion the team captain is shouting at me to hurry up. My gaze wanders around the shed in search of white-soled boots but they are nowhere to be seen. Instead I spot a pair of boots hidden underneath the changing bench. They are dirty and wet and slightly the worse for wear. In desperation I sneak them away and find that they fit comfortably. When the game is over I put the stolen boots into my locker and keep them for future use. But as the weeks go by my conscience is troubling me for taking some other boy's boots and when I go to confession the priest tells me to make retribution by putting a shilling in the church poor box every week. About a month later I'm watching a match between the older boys and notice that the prime bully of Year 2 is wearing white-soled boots. They look like mine but I have no proof and even with proof, I dare not challenge him. I continue paying my money to the poor box and keep the story of the boots a secret from Sara.

On the last Sunday of each month Mammy would come to see us and would drive out along the Muckross road. She had discovered a pleasant, quiet spot, where she would park the Morris Minor and open up the picnic box she had brought. She would also have a bag of goodies for both of us and two large pots of jam, with our names clearly labelled. Jam was not on the menu in the refectory but there was a large cupboard where we could store the jam. Every second month she would bring the hair clippers and give both of us a quick trim. For some reason or other we preferred to do it this way rather than go to the barber who came to the college once a week. Daddy came with her on a few occasions and treated us to lemonade in the snug of a local public house. On other occasions Tom or Dan would be there telling us everything that was happening at home and making me feel homesick. On returning to the college I would feel a dark depression set in that would last through the evening and the night. I would seek out the hidden places to be by myself to avoid the thoughtless jibes and longed to be in my bed asleep. By Monday morning the feeling would have passed. I was back to myself again and ready to slot into the usual routine.

As first term was drawing to a close, I was studying hard and had a good grasp of all subjects, but I never got to sit the exams. Auntie Lizzie had died suddenly a week before Christmas and Kevin and I were granted early leave to attend her funeral. I was thrilled to have escaped the exams but it was to have unforeseen consequences. Because Mammy was busy with the funeral preparations a neighbour came to collect us. We came home to a sombre house. Lizzie was waked at home as was the custom. Mick had vacated the downstairs bedroom and this was where Lizzie was laid out. The sister he couldn't love in life lay peaceful in his bed. Death brought to an end the animosity they shared in life, but it would weigh heavily on Mick's mind. He sat by the table, looking pale and drawn, as he talked to some of the mourners in the kitchen.

The house was thronged with people and reeking of tobacco smoke. The parlour was full, the kitchen was full and the bedroom, where Lizzie was laid out, had ten or twelve women sitting on chairs around the bed. Some wore the traditional black shawls that covered their heads and shoulders. They spoke in hushed tones and looked me up and down as I approached the death bed. Lizzie looked calm and serene and her pallid face had lost all wrinkles. Her elegant fingers were clasping a white rosary-bead to her chest. I thought about those fingers and the care they had given me long ago, by the fire in the upstairs bedroom. I mouthed a silent goodbye to Lizzie and turned to go. As I was leaving, an elderly lady asked: "Which of the boys are you" and held on to my hand for longer than was comfortable. A box of snuff was being passed around and some of the women had yellow stains on their nostrils. I felt awkward in these strange surroundings and made my exit as quickly as possible. Down in the kitchen, Mammy and Sara were moving about, offering tea and sandwiches to the women while Daddy was distributing Guinness or whiskey to the men. Shortly, the hearse arrived, the rosary was said, and Lizzie was taken away to Lixnaw church.

It was a sad occasion, things would never be the same with Lizzie gone but it was so wonderful to be at home and Christmas was coming. Little did we know then, as we tried to enjoy a dampened-down festive season, how Lizzie's sudden death would affect our futures. From the outset, she had volunteered her pension to help pay for college fees and now that source of income was gone. Money was scarce and hard decisions had to be made. Keeping two of us boarding in St. Brendan's was no longer feasible, so Kevin, who was seen as the one who would inherit, was withdrawn from college at the end of the year to work the farm. Instead of his liberal classical education he would travel to Tralee to do an agricultural course. Lizzie's death had a knock-on effect on my education also, but in a very different way.

A meritocratic system of seat allocation operated in each classroom in St. Brendan's, dependent on how students performed in the exams. Not having taken the exams at Christmas, I was placed at the back of the classroom for second term, sitting beside the boy who had the distinction of coming last. I had been in the front row and had always been very conscientious. Now I noticed that some teachers never focused on the back row and I felt left out and disconnected from what was going on. I found it difficult to pay attention and often had difficulty in seeing the blackboard or hearing what was being said. The slippage began to set in almost immediately. My desk mate liked to draw funny pictures and write amusing slogans for our mutual entertainment and so we both wasted much of our time.

On one particular day the when the history teacher was missing we were given instructions to study the next chapter on our own, unsupervised. Half way through the class my doodling buddy insisted on sharing a joke with me; a brief interlude of three or four minutes at most. We were both in fits of giggles just as the college president was spying on the class through the back window. The following morning at breakfast in the long refectory he summoned us both to the duty podium. We were to be made an example of for wasting study time. Silence descended as he admonished the two errant youths, wagging a long thin forefinger. Then grabbing each of us by the hair, he shook our heads violently and in a panting voice declared that he was sending for our parents.

Two days later I'm summoned to the luxury front reception room of the college where a pale-faced Mammy awaits me. I can hear him lecturing her as I enter the room and he's saying that she should reprimand me for not taking my study seriously. His eyes are sharp and piercing and I can see that Mammy is intimidated. She nods her head but makes no other reply. Without looking at me he turns on his heel and exits the room, closing the door behind him. I stand there ashen faced not knowing what to say. A moment of silence hangs heavily between us. I feel sorry that Mammy has to endure all this anxiety and humiliation on my account. She had dropped everything at home as soon as the urgent letter arrived and had driven to Killarney in a state of panic, not knowing what to expect. She looks at me with a bewildered expression on her face but not a cross word escapes her lips. I mutter 'Sorry Mammy' and try to explain that it was only a momentary lapse into the world of humour. Awkwardly she gives me a hug and tells me she will see me, as usual, at the end of the month. We say goodbye; she goes one way and I the other, back to my classroom where I hope I don't have to explain my absence to the class teacher.

But a greater impediment to my academic performance lay ahead and was to disrupt my college life from the age of fifteen onwards and seriously impact on exam results from then on. By year three the indoctrination had kicked in. Each year the annual retreat imposed a three-day silence in which prayer, meditation and high level indoctrination filled all our waking hours. Sexual morality was top of the agenda and the constant drip-drip of indoctrination had filled my head with obsessive doubt, worry and guilt feelings. Scruples began to take hold. The Oxford dictionary defines scruples as "an uneasy feeling arising from conscience that tends to hinder action" and the Catholic Encyclopaedia defines it as "an unwarranted fear that something is a sin, which, as a matter of fact, is not". We were all in our adolescent years with raging hormones and a never ending interest in sexual matters. But sex instruction was taboo and was only gleaned through the negative perspective of how sinful all sexual matters seemed to be. Some of the Passionist priests who came to deliver the retreats knew how to turn the indoctrination thumbscrews. It seemed to me that only a de-sexualised human being was pleasing to God. Their main focus was on bad thoughts, words and deeds. "Were you entertaining impure thoughts?" was a constant question in the confessional. A devil-may-care friend, nicknamed Randy, joked with me after confession that he wasn't entertaining impure thoughts but was entertained *by* them. I wished I could be that carefree and envied his peace of mind.

Thoughts were difficult to control but it was explained that a sexual thought was okay if you didn't take pleasure in it, but my fevered brain was never quite sure whether I did or not. Avoidance of impure thoughts became a continuous obsession that greatly affected my study. I would waste my study time agonising over whether or not I was guilty and if guilty, whether or not it was a venial sin or a mortal sin. At weekly confession on Saturday the priest would ask "How many times?", so keeping a head-count of sins was a constant requirement. One evening Randy had found a magazine with a photograph of a beautiful woman whose long blonde hair only partially covered her naked breasts. I had looked at it, but had my gaze lingered long enough for it to be a sin? I was like Hamlet; *did I or didn't I, that is the question.* Had I taken pleasure in it? "Honestly, Your Honour, I don't know". And so through the week I agonised but Saturday's confession made it clear; the magazine was 'an occasion of sin' and my classmate was a 'bad companion' who should be avoided.

Scruples became most acute before receiving Holy Communion in the College oratory every morning. I knelt through the Mass in an agitated state of mind debating with myself whether or not I was in the state of grace. But as every other student was going for communion, how could I remain behind with everyone looking at me. But receiving communion with a sin on my soul was flying in the face of God and so the agitation and deliberation continued unceasingly. The sword of Damocles hung over my head every waking hour; if I were to die in the state of sin eternal damnation would be my fate. The College oratory was a beautiful place, quiet and peaceful and I would go there at night trying to explain myself to Jesus and apologising profusely for my unworthiness. One priest had told us that with every impure thought we took up that Roman hammer and smashed it into the hand of Jesus. Other students felt I had a vocation and was destined for Maynooth because I spent so much time in the oratory. I didn't or couldn't confide in anybody and just struggled along in my own secret little hell for many years.

In college oratory, a soft-spoken Passionist priest had explained in detail what scruples were, but knowing about the condition didn't help. There were many times when I was totally lost to myself and to those around me and couldn't snap out of it. I used to wonder if other students were experiencing the same but was too embarrassed to confide in anyone. In his many long sermons the word pleasure seemed to mean sexual pleasure only and had taken on bad connotations. Going to confession during his retreat was compulsory because he had a mechanical device for keeping a head count. During confession he seemed obsessed with penile erection as a threat to my immortal soul. In the privacy of the dark confessional, he advised the wearing of swimming togs when taking a bath and when in bed to wrap a rosary beads around 'the thing between your legs' - his words, not mine. The lid has been lifted on physical sexual abuse but this too constituted abuse, abuse of vulnerable young minds, even though we must assume it was conscience-driven on his part. I certainly could have done without it; it wasted so much of my precious time and it took many years to break free.

Retreats came and went but low-level anxiety remained as my constant companion. It began when I woke in the morning and was there when I tried to sleep at night. But physical activity would bring temporary relief, whenever it was possible to be active. I could lose myself in whatever game I was playing and close down the analysis. The college had four magnificent handball alleys and I loved handball. Randy always had a good alley cracker and we would play till our hands were ready to drop off. On Saturdays when we went to confession in the cathedral he would bring the ball in his pocket. St. Mary's Cathedral is pleasantly situated with a spacious lawn on either side, so after confessions we would go to opposite sides of the cathedral and throw the ball to each other over the high vaulted roof, back and forth many times. On one warm Saturday, two of the dayboys from our class were walking by. They were carrying swimwear and towels in a carrier bag and told us they were heading for the lake to go swimming and asked if we would join them. But, as boarders, we were forbidden to leave the college grounds without permission. After a few minutes hesitation, we decided to chance it. It was risky, but this was too good an opportunity to pass up. They waited for us while we collected our gear and then guided us through woodland and fields until we came to the lake's edge. It was perfect; a little secluded curved beach with the lake water lapping on the pebbly sand. I couldn't swim, even though Mikie had tried to teach me in the sea at Meenogahane, but the local boys assured me that the lake was very shallow at this point. I dog-paddled for an hour in the lukewarm water and began to feel some buoyancy in my efforts. I was hooked and made immediate plans for the following Saturday. We did five Saturdays in all, five wonderful afternoons, and by then I had mastered the breast stroke and made two very good friends.

But my academic performance continued to suffer and, try as I might, I never succeeded in clawing my way back up the achievement ladder. I was struggling at Mathematics and when I reached the final year I was demoted into a class doing Pass Maths for the Leaving Cert. Around this time, I noticed that my vision was no longer as clear as it should be; myopia had set in, making things difficult in the classroom and in the hurling field. I think I was aware of my weakening sight for a long time but vanity prevented me from facing up to it. At Christmas, Mammy took me for an eye test to Florrie Carroll in Tralee. He selected a pair of nerdy-looking frames and said I would look "very distinguished". I hated the frames and I hated having to wear glasses. I wore them as little as possible and only in dire necessity and made do with my reduced vision. As luck would have it, Randy had been demoted into the Pass Maths class with me. We were still side by side in the classroom and he also needed glasses but didn't get any, so we shared my pair, alternately, to copy the teachers' writing from the blackboard.

Chapter Fourteen

In spite of my blurred vision, I was still reasonably good at hurling and held my place on the Crotta minor team. So too did John Shanahan, my classmate from Kilflynn school, who was also boarding in St. Brendan's. Whenever a game was coming up the club would request our release from Sunday morning study and somebody would come and collect us. On this particular Sunday it was Tim Lawlor and Brendan Twomey. They were waiting in the reception room where Mammy and I had endured the president's wrath. As the bad memories were surfacing, it was strangely comforting to feel their presence in this room. I felt they would have stood their ground and not have kowtowed as we did. But there was no time for such thoughts now, it was time to go. Both of them were talented players and, as we drove along the road, their enthusiasm for Crotta's success was infectious. It was just twelve o'clock when they dropped me home.

It was great to be home no matter how short the stay; home to Crotta where everything was homely and familiar, in spite of the sad bits. Poor Mick was sitting in the corner, lost to the world. I said "Hello Mick, it's me" but only got a blank stare in return. Daddy was working in the yard and I could see that his hip-joint was causing him pain. His body was leaning to one side as he walked, but his greeting, as always, was cheerful and welcoming. Sara was still the same and I gave her a quick hug before doing the rounds of the house, the farmyard and the fields, excited to see what changes had been brought about by Mother Nature or by manual labour. Diarmuid and Pat accompanied me. They were now nine and eight respectively and full of life and full of devilment. It was fun to be in their company. They brought home to me how much I missed being at home amidst all the things and places that were my treasured memories, where each season brought new challenges and a meaningful purpose to life. All my family lived here except me. I envied Tom taking the bus from home to the Christian Brother College in Tralee from Monday to Friday. The bus was his stomping ground and he told stories of having the *craic* (fun) with many pretty girls on the way home every evening.

Mammy had prepared the usual Sunday dinner of roast chicken but I was so nervous about the match I couldn't eat. After dinner Kevin drove us to Austin Stack Park in Tralee. In the dressing room it was good to see the faces I had grown up with and played hurling with in Stack's field. They were all there, looking fit and self-assured, while I was jittery with nerves. Todd, our star player, came by to say hello and talk for a minute and brought back some equanimity to my beating brain, but the gap between us had widened; our points of reference had somehow changed and the easy banter of long ago was difficult to recapture. A few words of advice from the trainer and we were ready for action. It was good to get out onto the field and release the tension. I played well that day and was still high on adrenalin as we were driven back to Killarney. Final term was nearing its end and the time was flying by at an alarming rate. I no longer dreaded going back into the college. The mournful sound of the cathedral bell, playing out its melody every fifteen minutes, no longer induced feelings of loneliness and depression. Knowing the end was near I had developed a new relationship with my surroundings. The memories they held, the experiences they had provided and the true friends they had generated, had suddenly become very precious to me. Even the bells had become a joyous sound.

It's the week before the exams, that final hurdle - the Leaving Cert. The weather is warm and bright sunshine blazes down from a cloudless sky. The college is quiet; all have gone home except those doing exams. I'm in the classroom cramming, trying to revise vast swathes of poetry, prose, history, geography, "Greek verse and Latin noun" and geometric formulae. Concentration is waning and the sun is beckoning. The call is too strong so I take some books with me to the playing field thinking I can study and enjoy the sun at the same time. The field is empty except for one boy, a boy who refused to worry, my devil-may-care friend. He's standing in front of the goalposts practising his hurling skills on his own. He asks me to join him but I say I'm trying to do some revision. He picks up a spare hurley and throws it in my direction. Resistance is futile. I pick up the hurley and run out to the halfway line and we engage in a 'long puck' contest. The day is so warm that we strip off to the waist to get a suntan and continue belting that ball backwards and forwards for hours.

My revision is suffering and worry is mounting. The following day, I have made a definite plan to spend the whole day 'swotting' focussing on two subjects. I get a good bit done in the morning, but in the afternoon a group are going for a swim at my favourite spot, so what can I do? We buy milk and biscuits in the local shop for our picnic and sneak off along the usual trail to the lake. In a nearby field a herd of red deer are grazing contentedly and I think how lovely for them to have no exams to worry about. The purple mountain opposite us is dotted with flowering rhododendron and its inverse image is reflected in the lake. The water is flat and calm like a sheet of ice and is surprisingly warm. With whoops and hollers we dash in and splash about for at least an hour, all worries forgotten.

The exams came and went and then it was all over. There was no grand finale, just a strange feeling of anti-climax. I said goodbye to those from faraway places, knowing that the likelihood of seeing them again was slim. It was one o'clock and I knew that the Morris Minor would be there to collect me, but I took a circuitous route from the exam hall for one last look around. I looked in at each of the four handball alleys and thanked them for the endless hours of fun they had provided. Across the lawn, the climbing roses on the tennis court fence were in full bloom and for the first time in my five years I realised what a beautiful picture they presented. A thought kept creeping into my head that I would miss all this. The uncharted waters of the future were looming large and were a daunting prospect. I knew I would miss the planned structure and the decision-less certainty that St. Brendan's offered. I walked around the lawn one last time and then headed for the dormitory. Everybody was gone and it felt eerily empty. My suitcase was packed and ready from the night before. I said "Goodbye bed, I won't be seeing you again". Kevin was waiting at the entrance gate to take me home.